MORE PRAISE FOR *THE ANIMALS' AGENDA*

"Thoroughly grounded in the history of animal protection, Marc Bekoff and Jessica Pierce have written a manifesto for a science of animal well-being, clearly distinguishing it from the science of animal welfare and articulating a vision for how this new science may inform a world in which humans live in community with their fellow creatures."
  —DALE JAMIESON, New York University, professor of
    environmental studies and philosophy; author of *Reason in
    a Dark Time* and coauthor of *Love in the Anthropocene*

"*The Animals' Agenda* upends the ways we grow up treating and demeaning nonhuman animals and their habitats. The well-documented style makes us want 'to think anew and act anew,' to use Abraham Lincoln's immortal words. Marc Bekoff and Jessica Pierce make you ponder and extend your own mind and sense of wonder about what the natural world means to us so we can in turn act more sensibly and kindly toward that sine qua non for human existence. *The Animals' Agenda* is a heartening book for all ages."
  —RALPH NADER, author of *Animal Envy: A Fable*

"This book is at the forefront of our evolving understanding, in science and ethics, of what we must do to protect animals in an increasingly human-dominated world."
  —DALE PETERSON, author of *The Moral Lives of Animals*
    and *Jane Goodall: The Woman Who Redefined Man*

"My eyes have been opened as never before, and if we long, as I hope we do, for a more balanced and kinder world, welcome this book, as I do, with open arms, minds, and hearts."
  —VIRGINIA MCKENNA, founder and trustee,
    The Born Free Foundation

"The call for a science of animal well-being is not only much needed but also long overdue."
—KIM STALLWOOD, author of *Growl: Life Lessons, Hard Truths, and Bold Strategies from an Animal Advocate*

"*The Animals' Agenda* is a bold and important book that everyone should read, argue over, and think about deeply. Not only does this book expose the obscene abuses of factory farms and lab animals, but it tackles deeper, harder, more complex questions as well. Should humans eat animals at all? Do we have the right to confine animals in zoos or aquariums? Is it morally defensible to let our cats roam free, where they will kill native wildlife, or to frustrate their hunting drive by restricting their freedom? People who love animals may come to different conclusions— but these are questions we should be discussing and debating if ever we are to evolve past the human-centered Anthropocene. *The Animals' Agenda* brings us closer to the day when our behavior toward our fellow species is determined not by convenience or greed but by compassion."
—SY MONTGOMERY, author of *The Soul of an Octopus* and *The Good Good Pig*

"*The Animals' Agenda* is a persuasively argued, lucid, and compassionate book that makes a powerful case for animal freedom. Marc Bekoff and Jessica Pierce highlight the crucial difference between animal welfare, which merely aims at minimizing animal suffering while keeping them in miserable conditions and eventually putting them to death, and animal well-being. There is no such thing as happy meat, a happy zoo, or a happy lab. Genuine animal well-being is the freedom from constraints imposed on animals' basic aspirations and natural behavior. This wonderful and inspiring book reminds us in particular that we should treat wild animals as our co-citizens on this planet and treat them with respect and protect their natural environment. A must-read by anyone who aspires to morality and justice."
—MATTHIEU RICARD, author of *A Plea for the Animals*

THE ANIMALS' AGENDA

The Animals' Agenda

Freedom, Compassion, and
Coexistence in the Human Age

Marc Bekoff and Jessica Pierce

BEACON PRESS, BOSTON

Beacon Press
Boston, Massachusetts
www.beacon.org

Beacon Press books
are published under the auspices of
the Unitarian Universalist Association of Congregations.

20 19 18 17   8 7 6 5 4 3 2 1

This book is printed on acid-free paper that meets the uncoated paper
ANSI/NISO specifications for permanence as revised in 1992.

Text design and composition by Wilsted & Taylor Publishing Services

Cover image, "The Babysitter," © Thomas D. Mangelsen, Images of Nature,
http://mangelsen.com

LIBRARY OF CONGRESS CATALOGING-IN-PUBLICATION DATA
Names: Bekoff, Marc, author. | Pierce, Jessica, author.
Title: The animals' agenda : freedom, compassion, and coexistence in the human
    age / Marc Bekoff and Jessica Pierce.
Description: Boston, Massachusetts : Beacon Press, [2017] | Includes bibliographi-
    cal references and index.
Identifiers: LCCN 2016021745 (print) | LCCN 2016046473 (ebook) | ISBN
    9780807045206 (hardcover : alk. paper) | ISBN 9780807045213 (e-book)
Subjects: LCSH: Animal rights. | Animal welfare—Moral and ethical aspects. |
    Human-animal relationships—Moral and ethical aspects. | Liberty.
Classification: LCC HV4708 .P566 2017 (print) | LCC HV4708 (ebook) | DDC
    179/.3—dc23
LC record available at https://lccn.loc.gov/2016021745

On the open plain under a lone acacia tree, a small pride of lions—an older female with three younger members—protects a precious eight-week-old cub. Each lioness takes on different roles depending on her age and abilities to help ensure the survival of the pride. The cub's mother, the most powerful and experienced hunter, focuses her attention on the passing wildebeest migration while one of the "aunties" concentrates her energy on babysitting the very precocious and last remaining treasure of the pride.

Jessica dedicates this book to Sage,
who embodies a future of compassion.

Marc dedicates this book to his wonderful
and compassionate parents, who always
stressed the importance of freedom and
supported him as he pursued his dreams
of trying to save the world, although at times
they weren't quite sure where he was heading.

Contents

CHAPTER I

Freedom, Compassion, and Coexistence in the Human Age

There comes a time when one must take a position
that is neither safe nor politic nor popular, but he
must take it because conscience tells him it is right.
—Martin Luther King Jr.

News headlines these days often center on animals. Stories seem increasingly to be of two types. The first involves reporting on what might be characterized as "the inner lives of animals." Scientists regularly publish new findings on animal cognition or emotion, and these quickly make their way into the popular press. Here is a sampling of some recent headlines:

PIGS POSSESS COMPLEX ETHOLOGICAL TRAITS SIMILAR
TO DOGS AND CHIMPANZEES

SQUIRRELS CAN BE DECEPTIVE

CHICKENS ARE SMART, AND THEY UNDERSTAND THEIR
WORLD

RATS WILL SAVE THEIR FRIENDS FROM DROWNING . . . NEW
FINDING SUGGESTS THAT THESE RODENTS FEEL EMPATHY

NEW CALEDONIAN CROWS SHOW STRONG EVIDENCE OF
SOCIAL LEARNING

ELEPHANTS GET POST-TRAUMATIC STRESS TOO: CALVES
ORPHANED BY THE KILLING OF THEIR PARENTS ARE
HAUNTED BY GRIEF DECADES LATER

FISH DETERMINE SOCIAL STATUS USING ADVANCED
COGNITIVE SKILLS[1]

The other type of news story focuses on individual animals or a particular group of animals who have been wronged by humans in some significant way. These stories often create a social media frenzy, generating both moral outrage and soul-searching. In particular, these stories highlight instances in which the freedom of an animal has been profoundly violated by humans. Some of these recent hot-button stories include the killing of an African lion named Cecil by an American dentist wanting a trophy head; the killing of a mother grizzly bear named Blaze, who attacked a hiker in Yellowstone National Park; the case of a male polar bear named Andy who was suffocating and starving because of an overly tight radio collar placed around his neck by a researcher; the "euthanizing" and public dissection of a giraffe named Marius at the Copenhagen Zoo because he was not good breeding stock; the ongoing legal battle to assign legal personhood to two research chimpanzees, Leo and Hercules; the exposure of SeaWorld for cruel treatment of orcas, inspired by the tragic story of Tilikum and the documentary *Blackfish*; and the killing of a gorilla named Harambe at the Cincinnati Zoo, after a small boy fell into the animal's enclosure. The fact that these events have created such a stir suggests that we are at a tipping point. People who have never really been active in defense of animals are outraged by the senseless violation of these animals' lives and freedom. The growing awareness of animal cognition and emotion has enabled a shift in perspective. People are sick and tired of all the abuse. Animals are sick and tired of it, too.

The Importance of Freedom

Freedom is one of the values we cherish most. Broadly understood, we are free if we are not imprisoned or enslaved, and not unduly coerced or constrained in our choices or actions. Freedom can be difficult to define, but we know when we lose it or when it has been taken from us. Human rights organizations are appropriately concerned when certain groups of people are exploited for their labor, like migrant workers forced into virtual slavery on fishing vessels or toiling in fields for little pay. They are concerned when groups of people are exploited for their bodies, as when young girls are forced into the sex trade. And they are concerned when groups of people are not allowed to move about or speak freely or engage in cultural rituals that are important to them. We also value the freedom to choose our family and friends, to bear and raise children, to think for ourselves, and to work for a decent living. Of course, there is no such thing as pure, unadulterated freedom—we are subject to our unconscious impulses, genetics, unspoken social conventions, and government rules that ensure public safety and order. But we are nonetheless free in important respects. Some measure of freedom is fundamental to human well-being: it provides the substrate for human flourishing.

Yet although we prize our freedom above all else, we routinely deny freedom to nonhuman animals (hereafter, animals) with whom we share our planet. We imprison and enslave animals, we exploit them for their labor and their skin and bodies, we restrict what they can do and with whom they can interact. We don't let them choose their family or friends, we decide for them when and if and with whom they mate and bear offspring, and often take their children away at birth. We control their movements, their behaviors, their social interactions, while bending them to our will or to our self-serving economic agenda. The justification, if any is given, is that they are lesser creatures, they are not like us, and by implication they are neither as valuable nor as good as we are. We insist that as creatures vastly different from us, they experience the world differently than we do and value different things.

But, in fact, they *are* like us in many ways; indeed, our basic physical and psychological needs are pretty much the same. Like us, they want and need food, water, air, sleep. They need shelter and safety from physical and psychological threats, and an environment they can control. And like us, they have what might be called higher-order needs, such as the need to exercise control over their lives, make choices, do meaningful work, form meaningful relationships with others, and engage in forms of play and creativity. Some measure of freedom is fundamental to satisfying these higher-order needs, and provides a necessary substrate for individuals to thrive and to look forward to a new day.

Freedom is the key to many aspects of animal well-being. And lack of freedom is at the root of many of the miseries we intentionally and unintentionally inflict on animals under our "care"—whether they suffer from physical or social isolation, or from being unable to move freely about their world and engage the various senses and capacities for which they are so exquisitely evolved. To do better in our responsibilities toward animals, we must do what we can to make their freedoms the fundamental needs we promote and protect, even when it means giving those needs priority over some of our own wants.

The Five Freedoms

Many people who have taken an interest in issues of animal protection are familiar with the Five Freedoms. The Five Freedoms originated in the early 1960s in an eighty-five-page British government study, *Report of the Technical Committee to Enquire into the Welfare of Animals Kept Under Intensive Livestock Husbandry Systems.* This document, informally known as the Brambell Report, was a response to public outcry over the abusive treatment of animals within agricultural settings. Ruth Harrison's 1964 book *Animal Machines* brought readers inside the walls of the newly developing industrialized farming systems in the United Kingdom, what we have come to know as "factory farms." Harrison, a Quaker and conscientious objector during World War II, described appalling practices like battery-cage systems for egg-laying hens and gestation crates for sows, and consumers were shocked by what was hidden behind closed doors.

To mollify the public, the UK government commissioned an investigation into livestock husbandry, led by Bangor University zoology professor Roger Brambell. The commission concluded that there were, indeed, grave ethical concerns with the treatment of animals in the food industry and that something must be done. In its initial report, the commission specified that animals should have the freedom to "stand up, lie down, turn around, groom themselves and stretch their limbs." These incredibly minimal requirements became known as the "freedoms," and represented the conditions the Brambell Commission felt were essential to animal welfare.

The commission also requested the formation of the Farm Animal Welfare Advisory Committee to monitor the UK farming industry. In 1979 the name of this organization was changed to the Farm Animal Welfare Council, and the freedoms were subsequently expanded into their current form. The Five Freedoms state that all animals under human care should have:

1. Freedom from hunger and thirst, by ready access to water and a diet to maintain health and vigor.
2. Freedom from discomfort, by providing an appropriate environment.
3. Freedom from pain, injury and disease, by prevention or rapid diagnosis and treatment.
4. Freedom to express normal behavior, by providing sufficient space, proper facilities and appropriate company of the animal's own kind.
5. Freedom from fear and distress, by ensuring conditions and treatment which avoid mental suffering.

The Five Freedoms have become a popular cornerstone of animal welfare in a number of countries. The Five Freedoms are now invoked in relationship not only to farmed animals but also to animals in research laboratories, zoos and aquariums, animal shelters, veterinary practice, and many other contexts of human use. The freedoms appear in nearly every book about animal welfare, can be found on nearly every website dedicated to food-animal or lab-animal welfare, form the basis of many animal welfare auditing programs, and are taught to many of those working in fields of animal husbandry.

The Five Freedoms have almost become shorthand for "what animals want and need." They provide, according to a current statement by the Farm Animal Welfare Council, a "logical and comprehensive framework for analysis of animal welfare." Pay attention to these, it seems, and you've done your due diligence as far as animal care is concerned. You can rest assured that the animals are doing just fine.

It's worth stopping for a moment to acknowledge just how forward thinking the Brambell Report really was. This was the 1960s and came on the heels of behaviorism, a school of thought that offered a mechanistic understanding of animals, and at a time when the notion that animals might experience pain was still just a superstition for many researchers and others working with animals. The Brambell Report not only acknowledged that animals experience pain, but also that they experience mental states and have rich emotional lives, and that making animals happy involves more than simply reducing sources of pain and suffering, but actually providing for positive, pleasurable experiences. These claims sound obvious to us now, but in the mid-1960s they were both novel and controversial.

It is hard to imagine that the crafters of the Five Freedoms failed to recognize the fundamental paradox: How can an animal in an abattoir or battery cage be free? Being fed and housed by your captor is not freedom; it is simply what your caregiver does to keep you alive. Indeed, the Five Freedoms are not really concerned with freedom per se, but rather with keeping animals under conditions of such profound deprivation that no honest person could possibly describe them as free. And this is entirely consistent with the development of the concept of animal welfare.

Welfare concerns generally focus on preventing or relieving suffering, and making sure animals are being well-fed and cared for, without questioning the underlying conditions of captivity or constraint that shape the very nature of their lives. We offer lip service to freedom, in talking about "cage-free chickens" and "naturalistic zoo enclosures." But real freedom for animals is the one value we don't want to acknowledge, because it would require a deep examination of our own behavior. It might mean we should change the way we treat and relate to animals, not just to make cages bigger or provide new enrichment activities to blunt the sharp edges of boredom and frustration, but to allow animals much more freedom in a wide array of venues.

The bottom line is that in the vast majority of our interactions with other animals, we are seriously and systematically constraining their freedom to mingle socially, roam about, eat, drink, sleep, pee, poop, have sex, make choices, play, relax, and get away from us. The use of the phrase "in the vast majority" might seem too extreme.

However, when you think about it, we are a force to be reckoned with not only in venues in which animals are used for food production, research, education, entertainment, and fashion, but globally; on land and in the air and water, human trespass into the lives of other animals is not subsiding. Indeed, it's increasing by leaps and bounds. This epoch, which is being called the Anthropocene, or Age of Humanity, is anything but humane. It rightfully could be called the Rage of Humanity.

We want to show how important it is to reflect on the concept of freedom in our discussions of animals. Throughout this book, we are going to examine the myriad ways in which animals under our care experience constraints on their freedom, and what these constraints mean in terms of actual physical and psychological health. Reams of scientific evidence, both behavioral observations and physiological markers, establish that animals have strongly negative reactions to losses of freedom.

One of the most important efforts we can make on behalf of animals is to explore the ways in which we undermine their freedom and then look to how we can provide them with more, not less, of what they really want and need.

Degrees of Freedom and Human Responsibility

Although freedom comes in shades of gray, our responsibility to take animal well-being seriously is black and white in that we must do the *very* best we can for *all* individual animals whose freedom is compromised by human activities. That is the basis of what we call the science of animal *well-being*, to be distinguished from the science of animal *welfare*.

In chapter 2 we present the basic building blocks of our argument. We explore the current state of knowledge of animal cognition and emotion, and explain how what we know strengthens a commitment to respecting animal freedoms. We also discuss exactly what animal welfare science tells us about the harms that billions of animals suffer under our care and why all the welfare-science research in the world won't make a significant difference for animals if we don't explore the more basic problems of captivity and constraint and losses of freedom.

We begin our Animal Freedoms tour in chapter 3, with the animals who suffer the greatest deprivations of freedom: agricultural "food" animals. These individuals live their entire short lives within the profound constraints of our industrial food system and are lucky to ever set foot on grass or feel the warmth of the sun or enjoy the company of their families and friends. In chapter 4 we move to a consideration of another clearly captive and highly constrained group: animals who are used in research and testing facilities around the world. In chapter 5 we consider a class of animals whose loss of freedom is more ambiguous and controversial: animals used in entertainment, particularly in zoos and aquariums, and in chapter 6, we explore the world of pet keeping and how our companion animals are more captive and far less free than we might realize. Finally, in chapter 7 we take a look at wild animals, discovering the surprising ways these animals are actually captive to human intrusions and manipulations, and are far less free than most imagine.

In the concluding chapter we return to the argument that what animals really want and need is more freedom. Although the plight of animals around the globe remains bleak, we may well be on the cusp of a revolution. The groundwork is being laid for new ways of thinking about animals and our relationships with them. Research on the cognitive and emotional lives of animals is helping reshape our ideas about who they are and what we can do for them, generating momentum for a significant paradigm shift.

We chose our book cover carefully, seeing a young wild lion on the Maasai Mara as a symbol of freedom. This cub seems, by his intent expression, to believe that the world is his domain and that he can do as he pleases. But his "babysitter" already seems to be telling him that although he's wild, he's not necessarily free.

The more we take animal freedoms seriously, the more we might feel obligated to change our behavior (don't eat that hamburger, don't buy that brand of toilet cleaner, don't take your kid to the zoo for her birthday, don't get a dog if you really don't have the time) . . . but these are simply choices we make every second of every day. These are things we may desire, but don't *need*. Many animals live impoverished lives because of our desires or our lack of awareness, and an impoverished life is a shame. Our purpose here is to highlight when, where, and how we compromise the freedoms of animals and how to undo these harms. We reproduce and consume as if we were the only show in town, and we need to recognize that it's not all about us.

While there are numerous people worldwide working extremely hard to right the countless and egregious wrongs to which billions of animals are subjected every day, the situation remains dire. It is time to evolve toward a science of animal well-being and move away from the science of animal welfare. The full expression of our humanity demands that we undertake a transition to the "Compassionocene," an era defined by our compassion for other animals. Expanding the meaning and application of the Five Freedoms—liberating them from the welfarist paradigm—will allow us to reassess what exactly it means to respect and enhance the freedom of animals.

CHAPTER 2

Can Science Save Animals?

The Five Freedoms offer an early and clear statement of what has become a strong current in many facets of animal protection: science will guide the way. The Five Freedoms brought into focus the idea that animals have social and physical needs that must be met in order that they have a reasonable quality of life, or what some call a "good life" or a "life worth living," and introduced the idea that science offers us insight into what these needs might be.[1]

An entire research program called "animal welfare science" has grown up around the idea that we can provide animals better welfare if we take an evidence- and data-driven approach to the billions of animals under human care, in order to prevent obvious and abject suffering and offer the most humane solutions to welfare problems.

There has been a strong feeling of hope that science can save animals. In earlier points in our careers, we both felt this sense of optimism. Yet for us and for many others working in animal protection, optimism has given way to disappointment and frustration. As time wears on, the promise that science will save animals has begun to dim. Despite the huge growth in knowledge, animals are in many ways worse off now than they were in the 1960s. It's reasonable to ask: How can this be?

Animal Cognition, Emotion, and Sentience:
The Phenomenal Growth of Research

Since the Brambell Report was published, an explosion of research into animal cognition and emotion has helped us elucidate in much greater detail what different animals need in different social and physical contexts. Indeed, the transformation in what we know and understand about animals is nothing short of revolutionary.

During the 1960s it was still scientifically acceptable to express skepticism about whether animals really can experience basic emotions such as fear or anger, and even to question whether animals can form friendships, experience a wide array of emotions such as joy, happiness, love, and depression, and feel pain. Today, such skepticism is rare. Talking openly during the 1960s about complex emotions such as grief or joy or personality in animals, as early researchers such as Jane Goodall did, was downright risky because it smacked of sentimentalism and went against the prevailing mechanistic view of animals. Now, research exploring personality differences among spiders, numerical abilities in chickens, empathy in mice, rats, and chickens, and optimism in pigs can launch a scientist's career.[2]

Consider, for example, the rapid and broad scientific acceptance of animal sentience. Simply put, sentience is the capacity to feel things, to have subjective experiences. During the 1960s sentience in animals was treated with considerable skepticism. Animal sentience throughout vertebrate taxa is now a well-accepted fact, and the focus of discussion has shifted to just how far, taxonomically, sentience might reach, with the answer being much further than anyone would have guessed.[3] For example, scientists have gathered evidence for sentience in octopuses, squids, crabs, reptiles, amphibians, and fishes.

As evidence of this shift, note the proliferation of governmental and institutional declarations asserting that animals are not insensate physical objects. Among these "Manifestos of the Obvious" we have, for example, a 2011 position statement published jointly by the American Veterinary Medical Association and the Federation of Veterinarians of Europe recognizing that "sentient animals are

capable of pain and suffering."⁴ The 2009 Treaty of Lisbon, signed by
European Union member states, mandated that, "in formulating and
implementing the Union's agriculture, fisheries, transport, internal
market, research and technological development and space policies,
the Union and the Member States shall, since animals are sentient
beings, pay full regard to the welfare requirements of animals."⁵ In
2015 France acknowledged sentience in pets and wild animals who
have been tamed or are being held in captivity. New Zealand gave
official support for animal sentience in June 2015. And as we write,
the province of Quebec is considering a bill declaring animals to be
sentient. We also have the Cambridge Declaration on Consciousness, which states:

> The absence of a neocortex does not appear to preclude an
> organism from experiencing affective states. Convergent evi-
> dence indicates that nonhuman animals have the neuroana-
> tomical, neurochemical, and neurophysiological substrates of
> conscious states along with the capacity to exhibit intentional
> behaviors. Consequently, the weight of evidence indicates
> that humans are not unique in possessing the neurological
> substrates that generate consciousness. Nonhuman animals,
> including all mammals and birds, and many other creatures, in-
> cluding octopuses, also possess these neurological substrates.⁶

These declarations give clear evidence that science is influencing
thought. Being a sentience denier is no longer a credible position.
The question of interest is not *if* animals are sentient, but rather *why*
sentience evolved and how far it might reach. This is a much more
interesting query and one that is consistent with scientific discover-
ies in a broad array of species.

Various capacities once thought to be uniquely human have been
discovered in other species: highly evolved systems of communication, tool use, self-awareness, culture, and morality. There is vigorous
debate about which specific cognitive and emotional capacities certain species possess, and a great deal of interest in the affective and
cognitive abilities and social behavior of animals once assumed to be

unfeeling, such as fishes, crustaceans, and insects. There is also now a broad acceptance of the basic Darwinian principle of evolutionary continuity, which recognizes that we share with other animals continuities not only in physical form, but also in a large repertoire of cognitive skills, patterns of behavior, social instincts, and emotions such as joy, pleasure, happiness, sadness, grief, and despair. Differences among species are differences in degree, not differences in kind, as Darwin noted more than 150 years ago.

The study of animal cognition and emotion has become a mature science, and continues to grow. In addition to an increasing catalog of books, there are now a large number of journals dedicated to understanding how animals think and feel—from well-established core journals such as *Animal Cognition, Animal Behaviour, Behaviour, Ethology, Journal of Comparative Psychology,* and *Animal Behavior and Cognition,* to the recently launched *Animal Sentience,* to a range of journals focused on the application of behavioral science within various industrial venues such as food production and laboratory research. An army of scientists and graduate students and "welfare professionals" now spans the globe. And yet . . .

Where Has All the Science Gone?

If you were to draw a curve representing the growth in our scientific understanding of animals, measured by the number of scientific articles including the keywords "animal emotion" or "animal cognition" or "sentience," you would see a sharp upward trend. From the 1960s to the present, the growth has been exponential, a steadily and sharply rising arc. One might expect that increasing knowledge about what animals are really like and what they need to be happy would be reflected in a pattern of increasing attention to these needs and a subsequent decline in practices that cause animals suffering and pain.

Not so. We are suffering from a problem social scientists call "knowledge translation": there is a large and widening gap between our knowledge base and the translation or application of this knowledge into policy and practice. A huge body of scientific literature is

available to guide our interpretive work with animals, but this knowledge is simply not being put to use in the service of animals. Philosopher Robert Jones at California State University, Chico, carefully
reviewed the past four decades of research into the physiological and
cognitive capacities of a wide range of species, and then compared
the state of scientific knowledge to animal welfare protections for
farm and lab animals. He concluded: "The moral status of animals
as reflected in almost all—even the most progressive—welfare policy
is far behind, is ignorant of, or arbitrarily disregards our current and
best science on sentience and cognition."[7]

Whereas we should see a downward trend in the numbers of
animals used in food production, research, and so forth, instead,
the number of animals involved is steadily climbing. The number
of animals killed for meat continues to grow, year by year, and not
by small amounts.[8] The use of animals within biomedical research
continues to expand, despite the increasing availability of alternatives and growing concerns that animal testing can deliver misleading data. Zoos and aquariums are as popular as ever. The number of
people keeping pets has increased dramatically, particularly those
keeping exotics; the number of animals "manufactured" and sold by
the pet industry is mushrooming. There has been a similar upward
trajectory in the number of wild animals culled by "management
professionals" and the number of species being driven to extinction
by climate change, habitat loss, and pollution. In each of these cases,
and likely in many others as well, the plight of animals seems to have
gotten worse, even as the science has gotten better.

These kinds of representations only show large trends, and don't
offer a nuanced picture of what's happening to animals and why. One
possibility is that although "uses" of animals such as pigs-to-pork and
mice-to-petri-dish show total numbers of animals involved continuing to climb, the actual lives of these many, many animals aren't
really so bad. We are able to provide some of these animals—more
than before—with certain pleasures and enrichments. Even if their
ultimate fate is an early death, their killing is humane and while they
are alive they are allowed to forage and scent-mark and otherwise
be normal animals. Perhaps conditions have vastly improved in the

past five decades so much that the rise in numbers of animals "used" has been offset by improvements in the quality of life these animals experience. So, science has made a difference to animals.

Another, more likely possibility is that in relative terms, animals are really, truly at a low point. We know so much about animals now that the way they are being treated is that much less morally defensible. You might excuse physiologist Claude Bernard for doing surgeries and dissections on live and unsedated animals, because he worked during the mid-nineteenth century, long before scientists had demonstrated that pain perception in animals is much the same as in humans. This knowledge wouldn't have made any difference to Bernard, because he considered his work of such profound importance to humankind. As physician Edward Berdoe, a contemporary of Bernard, wrote, describing Claude Bernard's attitude toward vivisection, "He does not hear the animals' cries of pain. He is blind to the blood that flows. He sees nothing but his idea, and organisms which conceal from him the secrets he is resolved to discover."[9] Nevertheless, Bernard's work makes some sense within the scientific milieu of his age. There may now be welfare guidelines insisting that researchers anesthetize animals before vivisection (unless there is a compelling "scientific" reason not to), but pain medications are uniformly underutilized or incorrectly used, so untreated pain in experimental animals is as much a problem as ever, yet morally offensive in the extreme, given that we know better and that effective remedies are readily at hand.

As an example of how science fails to translate into practice, consider the situation of the millions upon millions of mice and rats used in invasive laboratory research. These social rodents have a broad range of emotions very similar to our own. Both rats and mice are known to display empathy, and rats can detect pain in the facial expressions of other rats. More recently, research has shown that rats will readily seek to help a distressed cage-mate. In one experiment, rats even helped a rat in need instead of choosing to eat chocolate. Many models for understanding human depression and anxiety have been developed using rats, due to their close neurophysiological similarities to humans. Like humans, rats and mice exposed to repeated and prolonged stress fall into a state of despair. Yet despite clear evidence that mice and rats suffer negative emotions, we continue to use them in painful and distressing experimental protocols. Indeed, mice and rats aren't even considered animals under the Animal Welfare Act, which is patently absurd from a scientific point of view and demonstrates just how wide the knowledge-translation gap is.

Likewise, SeaWorld may have made sense when it opened its gates in 1964, when there was little awareness of the rich cognitive and emotional lives of marine mammals such as orcas, and the joy of watching them leap out of the water and splash visitors with their tailfins wasn't tempered by feelings of pity. Now that scientists have learned about the close and complex social bonds formed by these animals, we understand the horror of taking young orcas from their mothers and their pods. We also understand the profound deprivations imposed by captivity on an animal so intelligent, emotional, and far-ranging. Knowing what we do, this practice at SeaWorld now seems barbaric, and it is shocking that such places still exist in the twenty-first century.

When we started writing this book, we both conceptualized the gap between what scientists know about animals and how animals are treated in various venues and industries as a knowledge-translation gap. But the knowledge-translation gap only goes a short way in explaining why science isn't doing as much as it can to help animals. Clearly there is more going on.

Welfare Science: Trying to Use Science to Help Animals

"Animal welfare" has become all the rage. Entering the term into Google's Ngram Viewer, which charts the use of a word or phrase in a database of over five million books, you see the term really come into play during the 1960s, after which there is a sharp and steady spike in frequency, all the way up to the present. This reflects two trends: (1) greater public awareness of and attention to animal welfare issues, and (2) the growth of an entire scientific discipline dedicated to understanding and improving animal welfare. The Brambell Report was at once a response to the first trend and an incubator for the second.

The Five Freedoms, as the cornerstone of the Brambell Report and the welfarist agenda, state that we should seek to provide animals the basic goods that they want and need: food, water, space, health, comfort, and the ability to engage in at least a few normal species-specific behaviors. But what started to emerge during the research and writing of the Brambell Report is that what animals want and need isn't always clear. Even something as seemingly obvious as providing an animal appropriate housing might be devilishly complicated.

One of the most interesting pieces of the Brambell Report is found in an appended essay called "The Assessment of Pain and Distress in Animals," by noted ethologist William Thorpe. Thorpe raises a curious possibility: What if we could "ask" captive animals what kinds of environments they prefer? For the mid-1960s, this is a surprisingly forward-looking suggestion and one of the first formal proposals to scientifically study what animals want. And it turned out to be quite important, because many things that would seem like obvious welfare improvements—for example, that more space is always better—were not what animals themselves seemed to want.

One of the initial concerns that motivated Ruth Harrison and others to speak out about farm-animal welfare was the practice of confining egg-laying hens to what have become known as battery cages: long, stacked rows of very small and dense enclosures, with chickens confined one to each wire cubicle. The battery cages were typically constructed of wire mesh ("chicken wire") to which the anatomy of a chicken's foot is particularly ill-suited. Battery-housed hens suffer from painful foot deformities, broken bones, lameness, and open sores. The Brambell Report recommended that egg producers use a heavier metal mesh instead, but there really wasn't any empirical evidence that one type of flooring was better than another as far as the hens were concerned. So researchers B. O. Hughes and A. J. Black decided to "ask" the hens by giving them a choice of flooring. The hens actually preferred the chicken wire, most of the time, though they didn't show a strong preference.[10] The conclusion is that they showed a slight preference for the slightly lesser of two evils, but both options were unsuitable for their feet, so it really wasn't much of a choice.

It was the basic question of what animals themselves want that motivated and continues to motivate the discipline of animal welfare science, which as we have seen emerged in the wake of the Brambell Report. Early welfare-science discussions centered on the question of what will best satisfy the needs of animals, taking the Five Freedom needs as basic parameters (e.g., food, water, space, ability to engage in normal behaviors). The research has focused on trying to "ask" animals, by gaining a window onto their subjective experiences. In this sense, animal welfare science dovetailed perfectly with the emerging science of animal cognition and emotion, and the two have developed apace. Like the broader science of animal behavior, animal welfare science sought to explore what animals thought and what they felt, particularly within the context of animal agricultural systems, and later within other contexts in which poor welfare is routinely an issue.

Welfare Is About What Animals Feel

As leaders in the field of animal welfare science such as Marian Dawkins, Donald Broom, and Ian Duncan note, welfare is all about what animals feel, and you can get at feelings through carefully designed tests. Welfare science builds on the premise that we cannot directly access the inner state of animals, but that animal feelings have measurable correlates in behavior and physiology that we *can* observe, record, and measure. The questions researchers have generally asked follow the basic formula of the Five Freedoms: What do animals want to eat and when? What kind of shelter do they prefer? Is a particular experience causing them pain or distress? Studies have even tried to go beyond the relatively easy problems presented by the first four freedoms and have tackled the fifth and most difficult freedom, namely being able to exercise at least some normal species-specific behaviors, which can be exceedingly challenging, some might say impossible, to provide within a captive environment. Research might explore, for example, what kind of industrial-scale housing system allows hens to perform the greatest repertoire of hen-like behaviors, such as pecking, scratching, and dust bathing.

There is huge variety in the methods and objectives of welfare research, and we'll be writing about a lot of different studies throughout later chapters. But we'll give a few examples here. One of the most basic welfare experiments, often called "preference testing," simply offers animals a choice between two alternatives to determine which the animals prefer. A test might measure variables such as the amount of time spent in or with each option, when given free access. For example, the study by Hughes and Black on wire versus mesh flooring for battery-caged hens gave the hens a choice and measured how much time the hens spent on each type of flooring. The hens spent more time on the wire, and this was taken to mean that they preferred wire. Another early study by Marian Dawkins used preference testing to assess how much space Ross Ranger hens "need." Using hens housed four to a cage, she gave the groups a choice between small cages (247 square inches) and large (988 square inches). Imagine a rectangle 20 inches long and 10 inches wide—this is about the size of the small cage. She then recorded how long it took the hens to move into each cage, and determined that because they moved more quickly into the large cages, they preferred these.[11]

Knowing whether hens move more quickly into a large cage or spend more time on a plastic mesh floor or on a wire one doesn't tell us how much they care about the cage size or wire flooring. Perhaps they have just a very slight preference, or maybe one option makes a big difference to them. So researchers developed tests to determine how strongly motivated animals are by a particular choice. One way to study motivation is to make different choices have consequences. For example, animals might be confined after choosing their preferred option, or might be given an electric shock. The researchers could then see how much it takes to dissuade an animal from the choice. In a study of whether broiler chickens preferred plastic or mesh flooring, the experimenters added an "aversive." They illuminated the wire flooring at 800 lux. In this iteration, the chickens changed their preference from mesh to plastic flooring.[12]

Researchers have also used preference testing to assess how much animals like and desire improvements to their environment. For example, hens will consistently choose "furnished" accommodations, which have perches and nesting boxes, over a barren environment. Cage-enrichment preference has been studied in laboratory rats, too. Typically, an experiment will involve placing an enrichment item such as a wooden platform or some paper towels on one side of the cage and then recording how much time a rat spends on the "enriched" side of the cage and how much on the blank side. If more than half of the rat's time is spent on the enriched side, researchers conclude that rats have a preference for it.

Another line of research has focused on how rats rank these enrichments (e.g., they like paper towels better than platforms). And going yet a step further, some studies explore how various enrichments affect rats' behavior. For example, one study showed that in the presence of a nest box, some behaviors remained the same (eating, walking), while nest-box-related behaviors suddenly increased in frequency.[13] Imagine that!

More nuanced kinds of preference testing also have been developed, often drawing on techniques from within the field of economics and consumer behavior. For example, you can test the relative importance of different resources, asking whether a preferred food is more important than a preferred type of bedding, or how much better paper towels are than platforms. You can see whether there is inelastic demand for a given "commodity" by testing whether the animal still shows a preference when he is asked to work harder for it, where work might involve pulling a weight or pressing a lever repeatedly.

It is also possible to gain access to an animal's feelings by observing how an animal makes decisions. The underlying idea is that emotions influence the decision-making process, a fact that has been well established within human and animal psychology. Running this in reverse, you can "see" whether an animal is happy or unhappy based on what kinds of decisions they make. For example, a sheep who is depressed is more likely to view a novel object as something to be feared, rather than something to be investigated. This is called "cognitive-bias testing" and has been conducted in sheep, pigs, dogs, rats, chickens, starlings, humans, and a variety of other species, and has produced remarkably consistent results.

Cognitive bias can be an important indicator of welfare, because it reflects an animal's mood state. We know, for instance, that animals exposed to pain or fear over long periods develop a bad mood: they become chronically depressed. It may strike you as obvious that if an animal lives in a constant state of pain and fear, her welfare is compromised. And certainly much of welfare science is directed at proving the obvious, for example, "testing" whether taking young mammals away from their mothers is stressful for parent or offspring. But it is also possible that welfare may appear to be fine, but really isn't. Sometimes welfare can be compromised in ways that are not so obvious, for example, by the quality of lighting in a cage or by exposure to sounds that are inaudible to humans. And only by looking more deeply do we discover that the animals aren't really doing as well as we assumed.

Not only do researchers observe behavior to determine how animals feel about something, but they also often look at physiological markers (for instance, heart rate, blood pressure, and levels of cortisol) to decide whether an animal likes or dislikes something or is feeling good or bad. Negative feelings have measurable physiological effects, and physical effects are useful because they are objective and can be measured. (Welfare scientists really like to measure things.) For example, an animal under stress will have increased heart rate and increased levels of stress hormones such as glucocorticoids. As animal welfare scientist Ian Duncan says, "Measurements of impaired biological functioning, particularly those connected to decreased health and increased physiological stress responses, can provide good corroborating evidence that welfare is compromised."[14]

Looking at physiological data could tell us, for example, that laboratory mice are more stressed out when handled by male technicians than when handled by female technicians, a fact that would be highly relevant to researchers studying the effects of a certain drug or procedure on a group of mice because the differences in stress response could easily be a confounding factor that compromises data. A 2014 study of telomere length in African gray parrots offers another excellent example of this kind of research. Parrots who were housed in isolation had shorter telomeres than those who were housed with companions. Shortened telomeres are an indicator of stress. This kind of physical information is useful because, as Duncan says, it corroborates our suspicions that "welfare is compromised" and offers a more objective measurement of welfare than a behavioral choice between two options might.

Studies not only ask animals what they like the best, but also what causes them the most stress or pain. In fact, the vast majority of welfare research is focused on pain and suffering. "Preference" research, for example, is very often actually "aversion" research, focused on things animals don't like but that we are going to subject them to anyway. Research studies have measured animals' relative aversion to noise, vibration, and heat in various combinations. They have explored how aversive polluted air is, which euthanasia gases are most distressing to the mice who are being killed, and whether cattle are more stressed out by being yelled at or struck by handlers trying to move them through a slaughterhouse chute.[15] Scientists have studied whether rats exposed to an electric shock ("aversive stimuli") are more likely to prefer a familiar arm of a maze to a new arm. They have studied how much stress animals experience when exposed to overcrowding, small cages, incorrect social groupings or distribution of group members, and nearness of predators. In nearly all of these studies, the fact that a given activity is painful or stressful is already known. What researchers want to learn is how serious a welfare issue it is and, more importantly, whether it will compromise commercial productivity or experimental results.

The Development and Expansion
of Animal Welfare Science

The scientific study of animal welfare has developed in important directions over the past two decades. For one, the range of animals whose preferences we might explore has been expanded beyond the early focus of food and laboratory animals. Researchers have also begun exploring welfare considerations for zoo and circus animals, companion animals, and even wild animals. The science has also expanded beyond the negative experiences of animals to acknowledge the role of positive emotions and experiences and the need to provide animals happiness and not just protection from abject suffering. Even the Five Freedoms recognize that welfare goes far beyond lower-order interests (physical pain and pleasure) and affective states (fear, attachment, trust), and must give thought also to emotional and intellectual experiences, social relationships, creative expression; play; attention ("flow"); acquisition of knowledge and skill; and other higher-order interests such as happiness, a sense of fulfillment, and accomplishment.

Welfare science has been, and continues to be, extremely valuable, such as it is. Surely the lives of millions of animals have been improved based on animal welfare research. Yet stepping back from welfare science and placing it in a broader perspective allows us to see how we've shaped the research questions, and even to some extent their answers, to suit our own needs, and have all too frequently ignored what animals want and need, which is to not be held captive, not be used as a piece of lab equipment, not to be someone's Chick-fil-A. It helps us understand why good welfare is not and cannot be good enough. It also fortifies our belief that science, by itself, is not the answer to animals' problems, but rather can be a stumbling block to compassion.

Limitations of Welfarist Science:
Science Is Not Ethics-Free

As it matured, welfare science used techniques developed by scientists studying animal cognition and emotion and applied these to trying to improve the welfare of animals under conditions imposed on them by us. In other words, welfare science aimed to improve the lives of animals within the context of those situations in which human activities impose "welfare burdens" on our fellow creatures. Ian Duncan, for example, describes welfare science as a method "by which animals can be 'asked' what they feel about the conditions under which they are kept and the procedures to which they are subjected."[16] Welfare science may ask animals what they want. But the animals have actually been offered a very narrow range of questions and their answers are often ignored.

As we discussed above, welfare is about what animals feel. But it is about what they feel within the context of the status quo, and takes for granted that our ways of exploiting animals are morally sound. Indeed, welfare science has been instrumental in reinforcing the assumption that animals are here for the taking and that depriving them of the freedom to live their own lives is no problem, as long as we do it "scientifically and humanely." Going back to Brambell and the Five Freedoms, we can see this narrow, instrumental focus— that animals are here for us to use—already taking shape. Recall that the Brambell Report was written in response to concerns about the suffering of animals within the increasingly large-scale industrial farms that were replacing traditional methods of husbandry during the middle of the twentieth century. The practices that were under particular review were those used to raise laying hens, broiler chickens, pigs, and veal calves. At issue wasn't the question of whether or not to continue developing intensive food-production systems, but how to make modifications to the systems that would offset the suffering of animals to some degree. The Brambell Report essentially said: Given that intensive farming methods are now the norm, and farming is increasingly moving in this direction, what can we do to make the enterprise more acceptable to the public? The question was not, "Should we be intensively farming animals for human consumption?" or, "Is it ethical to breed, raise, and then slaughter animals in order to feed customer demand?" or even, "Are there ways to reduce human consumption of meat in order to reduce the total quantity of animal suffering and death?"

Welfare science is driven, first and foremost, by industry. Thus it can't really ask, "What do animals want and need?" It must, by necessity, focus on far more specific questions. For example, "What do animals in concentrated animal feeding operations (CAFOs) want and need?" Welfare science begins and ends within a "let us eat steak" mind-set. The underlying problems of profound exploitation and suffering of animals are swept aside, like dirty shavings in a pigpen.

Marian Dawkins's *Why Animals Matter* is a perfect example of welfare as a justification tool. She spends an entire book talking about what animals need and want, and how to measure their welfare, but fails to mention the most basic things they want and need: *freedom from captivity and from human exploitation*. This is not to point fingers at Dawkins. If you look through the entire corpus of books on animal welfare science, you will find no discussion of Freedom with a capital *F*. You will find little to nothing that challenges the status quo, or that raises serious moral objections to the various ways in which we routinely compromise animal welfare. That welfare will be compromised is a given.

While the desire of welfare science to reduce suffering has certainly been genuine, pragmatic reasons for seeking better welfare have provided the primary impetus for the Five Freedoms and for welfare science. Because many people are moved by abject suffering of animals, intensive farming chafed against the public conscience and some kind of response was needed. Additionally, producers were discovering that there is a point of diminishing returns, where compromises to animal welfare begin to decrease the productivity of a farming system. Too many animals die, or become so stressed out that they stop laying eggs or stop getting meatier, and profits start to fall. Working to improve welfare conditions could be a win-win for the animals and their abusers.

The motivations for animal welfare science are not, then, as noble as they might at first appear. There is good reason to want dairy cows to have decent welfare: they produce more and better milk. We want chickens to be "happier" so that they don't engage in as much destructive infighting, which costs producers profit. Laboratory researchers don't want stressed animals, because stress causes physiological changes that can contaminate data; zoo managers want animals to survive (they are costly to buy), and also want animals to be fruitful and multiply so that offspring can be sold or traded. Zoo visitors want to see "happy" and well-functioning animals.

An inherent tension has dogged the science of welfare, particularly as it has drawn data, methods, and inspiration from the broader scientific study of animal behavior: it is impelled to ignore its own conclusions. Welfare science is predicated on the fact that animals have subjective experiences, yet many in the field downplay these very same subjective feelings when it comes to implementing welfare guidelines or drawing moral conclusions from their work.

Dawkins's work embodies this tension. She insists that animal welfare science is all about what animals feel and that we can study animal "feelings" empirically, albeit indirectly. She also maintains that we cannot know for certain whether animals are really conscious and cannot be sure they have feelings. Using the same data as the drafters of the Cambridge Declaration on Consciousness, she somehow concludes, "There is no proof either way about animal consciousness and that it does not serve animals well to claim that there is."[17] This "agnosticism," as she calls it, is essential to Dawkins's larger welfarist agenda, which is to defend our current practices by arguing that animals have value only in their utility to humans. Dawkins, while being an important advocate for animal welfare, stabs animals in the metaphorical back, because at the same time she denies that we can know for certain whether animals are really conscious. This move leaves the door open for sentience deniers, who can claim that we don't have proof because scientists disagree. This is like climate deniers saying that climate change doesn't exist because scientists disagree about some of the fine details of how our planetary catastrophe is likely to unfold. Dawkins claims, "to make sure animal welfare stays on the agenda, we need to focus on the argument that animals provide a service to humans rather than that animals are conscious, intelligent beings." She goes on to say, "Animals matter because they are useful to us."[18]

The fact that welfare assessments are "science-based" gives them a patina of acceptability. Yet it must be kept in the forefront of our attention that science is not and never can be completely objective. Consider the concerns raised by animal researchers Joy Mench and Janice Swanson in "Developing Science-Based Animal Welfare Guidelines."[19] Value judgments are always a part of scientific assessments, they say. You collect numbers and data and there is some rigor to it, but then you have to decide what these mean to us and to animals. Mench and Swanson use the example of a seventy-two-square-inch minimum space recommendation for hens. Seventy-two square inches is not a scientific answer to the question "How much space do hens want and need?" It is a value judgment based on body measurements (how much space do hens need to spread their wings?), observations of how much space hens need to stand and lie down, mortality rates of hens in different-sized enclosures, and individual hen egg production. They write: "A different decision about the minimum recommendation would have been reached had the committee given more weight to preference testing and use of space studies, since these indicate that hens need and want more space than 72 square inches."[20]

We formulate questions to serve our purposes, gather information based largely on what answer we're looking for, and call it "welfare science." Science is not value-neutral, and as much as animal welfare scientists seek to approach their studies "objectively" and from a rigorously evidence-based stance, the entire enterprise is suffused with value assumptions about who animals are in relation to humans, and how human needs and desires take unquestioned precedence over the needs and desires of the animals themselves.

It is curious that the Brambell Commission used the language "Five Freedoms" in formulating their report, because only in the most jaded and cynical sense do their recommendations amount to "freedoms," especially since they are offered as ideals rather than actual standards. But the Five Freedoms perfectly encapsulate welfarist science: What can we do to "free" the animals whom we are exploiting? What can we do to "free" the hens and pigs in CAFOs (without actually getting rid of CAFOs)? The Five Freedoms are like a great big consolation prize: You, the animals, are clearly the loser of this game. But let's do what we can to make you feel better about it.

Getting back to our original question of why the science of animal cognition and emotion hasn't saved animals to the extent that it might, we can now formulate at least a partial explanation: the science of animal cognition and emotion has been largely co-opted by welfarism, which we define as the commitment to improving the lives of animals, within the status quo. Welfarism seeks to better the lives of animals within industrial agriculture, laboratory research, zoos, circuses and aquariums, and many other venues within which animals are currently being exploited. It may question particular practices within these industries—castrating animals without the benefit of anesthesia, for example, or keeping pregnant sows in tiny gestation crates—but the larger cultural and ethical commitments to using and abusing animals are left intact. Science hasn't transformed our behavior toward animals, because it is mired in an old but tenacious paradigm.

We are concerned that many welfare scientists are stuck within the welfarist paradigm and the social and cultural attitudes that are embedded within it. Indeed, when you see the phrase "animal welfare" you can pretty well bet that humans are doing something unpleasant to an animal or group of animals. "Welfare" means we are compromising something that animals want and need. Within the welfarist paradigm, we still deny animals the most basic freedom of all: the freedom to choose their own destiny.

The scientific endeavor itself may be part of the problem. To sum up, then, here is why science is failing to help animals as much as it could or should. First, science has tended to focus on different parts of the puzzle, like what kind of mesh causes the least severe injury to chickens' feet, while sidestepping the bigger picture—that putting chickens in cages is problematic. Second, science cannot help improve the lot of animals as long as it remains wedded to industry, because there is no internal motivation to translate knowledge about what animals need into significant changes in how we treat them. In fact, if anything, there is considerable pressure to maintain the status quo.

The Science of Animal Well-Being and a More Robust Concept of Freedom

We are in a critical time, when the welfare paradigm is growing increasingly unsteady under the weight of its own empirical conclusions and the expanding body of knowledge about animal cognition and emotion. The more we know about who animals are and what they want, the greater the welfare concerns become and the more difficult they are to satisfy. Welfare science could very well go from being a science that justifies an ugly status quo to being a transitional science, one that transforms how we relate to animals, as our moral paradigm catches up with our knowledge of animals and as we transition to a world of less violence. We can hope for more and more moments of disruption and discomfort, where paying attention to who animals really are allows us a greater measure of perspective, such as recognizing that pigs and rats are not so different from companion animals such as dogs. They are all mammals, and they all share the same brain structures and chemicals, and feel essentially the same range of emotions.

We are introducing "the science of animal well-being" because animal welfare science and the resultant welfare-based freedoms are too limited and can easily ignore the real needs and wants of individual animals. The science of animal well-being is not an expansion or refinement of the welfare paradigm; it represents a difference in kind rather than in degree. Figuring out what animals want is an act of interpretation, based on scientific knowledge and expertise; it can—and should—also be an act of advocacy. Attending to well-being means attending to real freedoms, including the psychological benefits for animals of choice, control, and agency.

The central argument of this book is that what we've learned about who animals are should, indeed must, make a difference in how we treat them. Yet as valuable as science is in furthering our understanding of animals, it is not going to make us more compassionate. Scientific evidence cannot produce obvious and agreed-upon moral "outputs." But it most certainly *should* inform our moral judgments and our behavior. Just as it is wrong, in our view, to deny the reality of climate science in shaping government policy, industry, and personal behavior, it is wrong to deny the reality of the science of animal cognition and emotion. The reality of climate change is, as Al Gore says, an inconvenient truth, because it urges us to move beyond our highly profitable reliance on fossil fuels and alter our daily lives, to give up things like conspicuous consumption. The truth of animal feelings is similarly inconvenient, in that it challenges our highly profitable animal industries and our personal habits.

Our agenda for the next five chapters is to explore what we know about the needs and wants of animals, including some examples of work done within animal welfare science as well as the much broader base of knowledge about the cognitive and emotional capacities of various species. We'll delve into some of the scientific limitations to preference and welfare studies, exploring the ways in which the kinds of questions asked have shaped and narrowed the kind of knowledge we have. In particular, we will look at how the well-being of animals is compromised in ways well beyond what a welfarist position would acknowledge. For example, welfare science addresses some of the ways in which we can make captivity more humane, but it typically fails to look more deeply at the costs of captivity itself.

Finally, we'll consider how we can begin to move from a welfarist to a well-being framework, what we can do to help bridge the knowledge-translation gap between welfare, well-being, and action. This is a significant paradigm shift, in which the interests of individual animals are not categorically subsumed under the needs and desires of humans, and where our moral responsibilities to animals themselves actually have some teeth.

CHAPTER 3

The Animals Whom We Eat

Tell people to stop killing dogs or cats and everybody
loves you. Tell people to stop torturing and killing cows,
chickens, pigs or turkeys and suddenly you're a self-righteous
asshole that needs to mind your own business.

Anonymous Tumblr post

The Stairway to Heaven

The storyline of Errol Morris's *First Person* documentary *Stairway to Heaven* unfolds through images, interrupted here and there by interview segments with iconic animal welfarist Dr. Temple Grandin, in which Grandin's voice punctuates the eerie background music. Blurry black-and-white video of passageways, of cattle moving forward in a group, are intermixed with complex architectural drawings of a slaughterhouse. The cattle make their way out of a crowded truck, where they were packed like sardines in a can, and are ushered through a series of pens. They move steadily forward through a curved chute and onto the killing floor. Curves, Grandin's voice explains in the background, are comforting to the animals because they think they are going home. The passageway guides the animals, now single-file, up a gently sloping ramp and onto a moving conveyor belt. A harness strapped around the belly lifts each animal, and they levitate into the hands of a plant worker who aims a bolt gun between their eyes and pulls the trigger, sending a metal rod crushing into their brain. This is what Grandin calls, with no irony, her "stairway to heaven." "Nature is harsh," Grandin says as she contemplates her innovation. "A modern slaughterhouse is much more gentle."

This new style of ramp designed for cattle on their way to the slaughter hold is perhaps the most famous innovation of the twenty-first century's most famous apologist for industrial farming.[1] Colorado

State University professor Temple Grandin's work has become synonymous with "farm" animal welfare. Grandin enjoys a position that is unique, in that she is hailed as a compassionate helper of animals while at the same time working within a venue in which billions of animals are harmed and killed. This makes Temple Grandin's work the perfect foil for understanding how welfarism goes astray. While Grandin has surely made contributions to factory-farmed animal welfare, she has also done more than anyone else to deflect attention from real freedom for animals. Even if a few animals are getting a "better life," it surely is not a good life.

Of all the venues of animal use we explore in this book, the food-animal setting imposes by far the most severe welfare problems, both in terms of the sheer numbers of animals involved and the nature and extent of the "welfare compromises" and pain we impose on them. No one denies that these animals are suffering—not those doing the welfare research and not even people working within the industry. A short excerpt from an article in the industry journal *Meat Science* on the welfare of "finishing pigs" offers a representative example. (A finishing pig, in case you don't know, is in her final life stage, when she will be fed and fattened with an eye to creating the tastiest bacon, pork, or processed product.) The slaughter, researchers note, is composed of a variety of stages, including transport, lairage, stunning, and sticking. "At each of these stages," they admit, "the animals are exposed to different stressors that, both individually and in interaction with one another, can compromise welfare."[2] This excerpt could have been drawn from hundreds of different industry publications, and could describe any number of different procedures or practices in which animals are "exposed to stress."

Welfare science originated within the context of agribusiness, and there has been extensive empirical research on the effects of intensive farming systems on animal well-being. We have more studies of the "preferences" of hens, cows, pigs, and sheep than of any other species, wild or domestic. In fact, we can use the quantity of welfare science in a given area as a rough marker of how severe the welfare compromises are and how hard people are trying to minimize the suffering to "acceptable" levels.

As with the other categories of animals we discuss in later chapters, there is great diversity in how animals fare within farming systems. We want to acknowledge up front that while some production systems involve complete violation of the freedoms of their animals and subject animals to horrible suffering, there are small-scale or family farms on which animals have a relatively good life, up until the time of slaughter. We'll talk at the end of the chapter about how some smaller farms are able to provide animals with greater well-being, but most of our focus in this chapter will be on large-scale industrial farming systems. Most of the animals who become someone's meal are "produced" in these large-scale systems. There has been steady growth since the mid-1960s in industrialized farms, and a large increase in the amount of factory-produced meat that consumers eat, in the United States and globally. Although we are focusing on the well-known and widely consumed pigs, cows, and chickens, it is worth remembering that many, many other animals are intensively raised for food, including rabbits, turkeys, geese, ostriches, emus, fishes, and crustaceans, to mention just a few.

Constraints on Farm Animal Freedoms: Confronting the Surreal

It may seem bizarre to talk about freedom in the context of food animals or to suggest that certain "more humane" alterations to their captive environment, like Temple Grandin's stairway to heaven, could possibly make them happy. Yet it is worthwhile exploring freedom in this realm, as surreal as it might seem. We'll see, for starters, that violations of the Five Freedoms are rampant; adherence to these animal welfare goals is not enforced and violations do not incur punishment but are, rather, fully accepted as the cost of doing business. Additionally, because the Five Freedoms are typically understood, as by the Brambell Report, as unachievable ideals, failure to achieve them is viewed as inevitable. What happens, then, is that welfare science focuses on minor improvements to caging systems or slaughterhouse design, without really examining the serious deprivations and constraints to freedom that our food-production systems, and our eating habits, impose on sentient creatures. We may proclaim that we should have the freedom to eat whatever we want, but this proclamation sounds mighty selfish in the context of a discussion of how profoundly we violate animals destined for our stomachs.

Animals in intensive-farming systems have essentially no freedom. They are confined to small cages or crates, or else they are packed into a large space with so many others of their kind that physical movement is highly constrained. Their biological development is controlled by us: they are genetically manipulated to develop in certain ways (nearly always physically deforming and painful) and given highly processed and regularized "feed" (to be distinguished from "food") to promote quick growth and fatten them up. They certainly don't have freedom to live a natural lifespan, as nearly all food animals are slaughtered while young, which may be a blessing.

In addition to physical constraints, food animals are unable, for a variety of reasons, to engage in normal behaviors, as individuals and as social beings. They have little to no control over social interactions and attachments. Either they are isolated, or they are housed in overcrowded groups that don't allow normal social interactions to take place and subject them to increased aggression from their fellow inmates. When densely housed, hens peck at each other and pull out feathers, while if allowed normal interactions they would typically only peck to maintain a social hierarchy. They rarely have a choice about where, when, whether, and with whom to reproduce, if at all. Most are denied the pleasure of sex, since the majority of breeding is accomplished artificially, and the parent-offspring bond is almost always broken, causing suffering for mother and baby. Maternal behaviors are thwarted. For example, dairy cows will bellow for their calves for days when separated, unable to nurse or lick their babies, and young animals cannot frolic and play.

Murder Most Fowl

The late Chris Evans, who taught at Macquarie University in Sydney, Australia, spent years studying chicken behavior and cognition with his wife, Linda. "Chickens," he noted, "exist in stable social groups. They can recognise each other by their facial features. They have 24 distinct cries that communicate a wealth of information to one other, including separate alarm calls . . . They are good at solving problems. As a trick at conferences I sometimes list these attributes, without mentioning chickens, and people think I'm talking about monkeys."[3]

Naturalist Sy Montgomery has also spent a lot of time getting to know chickens, and she is continuously amazed at their intelligence and personality. In addition to those mentioned by Evans, chickens have a number of other skills that may be surprising to those who haven't known any chickens personally. For example, they have the capacity for "object permanence," understanding that an object, when taken away or hidden, still exists. Chickens also, Montgomery learned, may use special "names" for the special humans in their life, according to a conversation Montgomery had with Melissa Caughey, author of A Kid's Guide to Keeping Chickens. Observing her flock of ten at her chicken compound, Caughey discovered that her chickens had invented a special name for her, which sounds like "ba-Ba-BA-BAA."

Montgomery recounts Caughey's story:

> "I was out there one morning, throwing scratch into the run," she told me, when she noticed her eldest, 6-year-old Oyster Cracker, "was talking to me in a different voice, one I'd never heard before." Oyster Cracker wasn't just uttering the greeting, "Brup? Brup?," a chicken hello. She certainly wasn't saying "bwah, bwah, bwah." (That's chicken for "I'm about to lay an egg." But Oyster Cracker had long since entered what Caughey calls "henopause"). Oyster Cracker was clearly saying, with increasing emphasis and tempo until the final, higher note, like trumpet fanfare, "ba-Ba-BA-BAA!"
>
> "It was quite regal-sounding," Caughey said, "almost like announcing the arrival of the Queen!" And then Caughey noticed other hens would say it, too—only when they first caught sight of her. Hence her conclusion: "When they see me, they call my name."[4]

Despite their possession of smarts and personality, chickens are routinely treated as though they have no brains and no feelings. Consider a few of the welfare problems faced by these remarkable birds. About 90 percent of commercially sold eggs come from hens housed in battery cages. In this system of "husbandry," each bird is confined to a wire cage so small they often cannot even fully extend their wings. Imagine having to live in a space so small that you could not straighten your legs. Your muscles would atrophy and you would be in constant pain. This is what happens to the hens. They have very little mobility or exercise, and they cannot forage, preen, or dust bathe, all behaviors that they are highly motivated to perform. Chicken feet are not designed for wire mesh; they evolved to perch on branches. In wire cages, their feet lose bone volume and strength, and they develop abnormal thickening of their toe pads, leading to a near-constant state of pain, a clear violation of the second and third of the Five Freedoms.

The fourth freedom is also routinely disregarded. Species-typical social behavior is totally ignored and disrupted; there are high levels of aggression, which is why de-beaking is "necessary"; subordinate hens cannot escape to a perch, as they might in a free-range setting. Forced molting (to encourage egg production) is induced by stressing the hens, either by depriving them of sleep, manipulating light/dark cycles, or by restricting food. About once a year, the entire population of laying hens is slaughtered and replaced. The so-called broilers, male chickens who will be slaughtered for their flesh, are physical mutants. They are genetically bred to grow huge muscles, especially in the breast, which their skeletal structure cannot support. They become crippled, and often completely immobile. Research has found lameness in 90 percent of broilers. Regrettably, humane slaughter laws don't apply to chickens or other poultry, even though they represent more than 98 percent of slaughtered land animals in the United States.[5] Slaughter methods are often obscenely cruel.

Chickens, like other animals, have behavioral needs. They have internally motivated behaviors they will experience regardless of where they are kept. This is one of the most empirically solid assumptions of animal welfare science, straight from the evolutionary-biology textbooks. It is also one of the keys to understanding why intensive-farming practices are so problematic for animals. Chickens, for example, will be motivated to nest, dust bathe, preen, and establish social hierarchies, no matter what environment they are placed in. They will have these behavioral needs even when they are housed in cages, even when the natural stimuli (e.g., a natural substrate to scratch out a nest or dust to bathe their feathers) are absent. They have an urge to scratch the ground in search of food, and will have the urge to forage even if they have a trough full of food.[6]

When chickens are kept in environments that inhibit or impair their natural behaviors, which they almost universally are on factory farms, they suffer physically and especially psychologically. This is why the Five Freedoms can be misleading: they suggest that if you provide animals with food, water, shelter, and some simulacrum of species-typical behavior, you have taken care of their needs. But an animal provided with food, water, and shelter can still suffer distress, a violation of the fifth freedom. Indeed, stereotypic behaviors, which are abnormal repetitive behaviors and a sign of psychological distress, are endemic among farm animals, a clear sign that all is not well.

Freedom to Eat

Let's focus for a moment on the first freedom—freedom from hunger. If an animal is provided sufficient food, then the first freedom, such as it is, has been satisfied. But does preventing an animal from experiencing physical hunger really resolve welfare concerns related to food? Next to breathing, eating is among the activities most essential to survival, and different species are exquisitely and finely adapted to meet their survival needs within the ecosystems in which they evolved. Many of the behavioral patterns for a given animal are directed at finding food, and animals are highly motivated to perform these food-acquiring behaviors, because these are so basic to survival. Providing a cow a trough of grain may satiate the physical hunger, but will not allow the cow to use any of the behavioral skills she has evolved to acquire food for herself. The behavioral urge to forage is present, even when cows are fed ad libitum. And this leads to welfare problems.

Oral stereotypies, in which an animal performs repetitive and seemingly functionless oral and oronasal activities, are prevalent in captive ungulates such as cows, pigs, and horses.[7] These behaviors include tongue rolling or object licking, bar biting, vacuum chewing (chewing when nothing is present) and sham biting, chain chewing or manipulation, and polydipsia (excessive drinking). Several significant causes of oral stereotypies are restriction of physical movement by bars or by tethering, early maternal separation, and barren environments. Another more subtle factor is related to food. When ungulates are held in captivity, particularly within intensive-farming systems, dietary needs and motivated foraging behaviors are both frustrated. Herbivorous animals such as ungulates would typically ingest large quantities of vegetation, since they need bulk to get sufficient nutrients and fiber, and would normally spend many hours in a day foraging and eating. But while a cow in pasture might spend an average of eight hours a day grazing, within an intensive-farming system she might spend only about twenty minutes eating. On a low-fiber diet, the cow will never feel satiated, and will be in a constant state of hunger, a clear violation of the first freedom.

Sometimes violations of the first freedom are built even more explicitly into the production systems. We mentioned already that chickens are induced to molt by extreme calorie restrictions. Another example of denying food to animals is in the care of pregnant sows, who are often placed on a restricted diet to control weight and to encourage maximal food intake during lactation. Because they are given only about half of what they would normally eat, they experience "prolonged high levels of frustrated feeding motivation."[8] Animals typically are fed "homogeneous foodstuffs" in a trough or manger. They have no choice in what they eat, cannot express preferences for different types of food as they would in a natural setting, and spend only a fraction of the time eating than they would under natural conditions. As humans well know, eating certain foods can be highly pleasurable, and what we'll next eat and when is often an issue of intense interest and anticipation. Yet animals in agribusiness settings, and most others in captivity, are generally offered food that is the same from day to day and which may meet a minimum threshold of palatability but is not actually very interesting or tasty. Eating is one of the great pleasures in life, yet we routinely deny this pleasure to food animals, the very animals whose flesh some people relish.

Recall the study we cited above in which chickens will work for their food, even when the food is offered for free. The same experiment has been conducted on a whole range of animals, and the results are the same: *animals don't necessarily want a free lunch.* Animals get pleasure from engaging in their natural feeding behaviors, because these behaviors are intrinsically rewarding. One of the trite comments often offered in defense of factory farming is that the animals are lucky because we feed them; they don't have to do any work and can just stand around all day doing nothing. But free lunch gets boring day after day. This isn't what animals want or need.

Hunger is certainly a welfare concern of top priority, and failing to provide sufficient food to a captive animal is cruel. Yet freedom from hunger doesn't really satisfy animals' needs. A more meaningful freedom would be the freedom to eat, including the freedom to procure one's own food, to eat the types of foods one has evolved to find most nourishing, and to search out (and hopefully find) foods that are highly pleasurable. Animals should be free to forage, scratch, hunt, root, and move around looking for the tastiest patch of grass.

What Food Animals Want and Need:
Preference ≠ Choice

The study of animal preferences really took root with research into farm animals like chickens, pigs, and cows. Even today, the body of literature on food-animal welfare dwarfs the literature on animals in other venues such as zoos or circuses or pet keeping. This may seem counterintuitive, but it actually makes good sense. The situations in which animal welfare is most severely compromised are those where we feel most compelled to respond to the suffering. Thus, as we noted earlier, the number and range of articles on a given practice can serve as a rough measure of the welfare problems associated with it. For example, a four-page report by the Humane Society of the United States on intensively confined food animals has 261 references on battery cages, gestation crates, and veal crates, and the listed references are just the tip of the iceberg for these welfare topics.[9] We can surmise, and we would be correct, that battery cages, gestation crates, and veal crates impose profound and egregious welfare compromises on animals.

As we pointed out in chapter 2, in the discussion of preference studies, "preference" doesn't really have anything to do with choice, at least in the context of most welfare science. This is truer in the realm of food animals than in any other area of animal-welfare-science research, for the obvious reason that the lives of food animals are so carefully and thoroughly manipulated by humans. So, "preference" really needs to be understood within a very limited context. For instance, Grandin's research on whether cattle have slightly lower cortisol levels in a curved chute than in a straight one has merely shown us that one option is slightly less distressing than the other. But the cattle really have no choice about what is happening to them. Likewise, research into the size of chicken battery cages—do chickens prefer seventy-two inches or fifty-eight inches?—has nothing to do with freedom or choice. Nor does this kind of "preference study" have anything to do with well-being, because both options are awful. This research is aimed at finding the least aversive conditions for the animals, but not really about finding out what they want or need or what is best for them.

Preference studies may result in somewhat less aversive conditions for animals, and this, perhaps, is better than nothing. The real value of welfare science, however, is in its attempt to understand the subjective experiences of animals, not by asking for a narrow this-or-that preference, but by really trying to understand how they experience the world. This is the revolutionary potential of welfare science, especially as it dovetails with the sciences of animal cognition and emotion.

The Cognitive and Emotional Lives of Food Animals

In 2008 Bill Crain and his wife, Ellen, opened Safe Haven Farm Sanctuary in Poughquag, New York. It provides a permanent home to animals rescued from slaughter and abuse, and the sanctuary currently houses eighty-nine animals, including chickens, goats, sheep, turkeys, partridges, pigs, and horses. All of these animals display their own kinds of intelligence and personalities, too numerous to describe, but we will let Bill's heartwarming story about Leo the Lucky Pig serve as an example.

One day a pot-bellied pig was brought to our farm sanctuary. Pot-bellied pigs are related to the standard breeds but are smaller. This pig was particularly small because he was only two months old.

All our staff members adored him, and he liked everyone. When people gave him a pat on the back, he rolled over for a belly rub. The word soon got out that there was a cute little pig at our farm, and visitors poured in. We named him Leo, after Leo Tolstoy, the Russian writer, pacifist, and vegetarian. Leo was brought to us by a family in Queens, who purchased him as a pet.

The family became very attached to Leo, and Leo became attached to them. But the family had to leave him alone during the day, and he sometimes cried. This disturbed the neighbors, who complained to the landlord. The landlord found out that Queens, like the rest of New York City, forbids residents to keep pigs, and he gave the family two weeks to get rid of him.

The family felt terrible. They didn't want to send Leo to an animal shelter, where he would probably be euthanized, or to a farm that would slaughter him when he came of age. (Pot-bellied pigs are often killed for food.) But, as the deadline approached, they couldn't find Leo a new home.

Our farm wasn't set up to care for a pig, but the situation was so urgent that we said we'd adopt him.

So the family put Leo in their car and drove him to our farm. The mother, however, stayed home. She felt the trip would be too emotional for her. After two months went by, she did decide to visit Leo at our farm, and she was thrilled to see him thriving. Leo seemed to remember her, too.

Although Leo is widely adored, he also gets into considerable mischief. He sneaks into chickens' aviaries and eats their food. He knocks over garbage cans in search of scraps. He crawls under gates and scampers around so vigorously that he frightens the turkeys and sheep. Staff members must constantly keep on the lookout for him. If he's not in sight, it's a good bet he's causing trouble.

The staff members do their best to control Leo's behavior, but it's difficult. If, for example, he's in an aviary eating chickens' food, it's difficult to get him to leave. The staff tries to shoo him out, but he ignores them. Occasionally a staff member resorts to picking him up and carrying him out, and he squeals loudly in anger.

We have had a few troublesome animals like Leo on our farm, and I have noticed that these animals soon respond surprisingly well to one or two particular individuals. In each case, the individual is exceptionally devoted to the animal.

For Leo, this person is Donna Scott. Whenever Donna calls, Leo comes to her. Even if he's eating other animals' food, he stops and runs to her.

I asked Donna, "How did you accomplish this?" She said, "When I see Leo, little hearts pop out of my head. I give Leo lots of rewards. Sometimes the reward is a small treat like a raisin, but most often the reward is just some form of affection. I always pet him or brush him when he responds to my call. I believe Leo can sense my love for him, I think all living beings respond to love."[10]

While Leo enjoyed stealing the chickens' food, Doink, a particularly clever adopted pig who lives with Joan and Don Hobbs at Happy Mama Acre outside Boulder, Colorado, is adept at stealing from two adopted alpacas, Tia and Junie B. Joan tells Marc that Doink is hardwired to know when she puts fresh hay in the alpacas' shed. Doink is very specific about it: he tips over the feeder, snuffles the hay flakes apart with his nose, and uses his hooves and snout to gather pieces that he then stuffs in his mouth. When Doink notices that Joan is there he takes a circuitous route back to his shed. Joan has a video of

Tia watching Doink steal her hay, in which she leans over and kisses him twice, as if telling him she is willing to share. But Doink seems to know that Joan does not approve. Joan proudly calls Doink "lovingly cantankerous."[11] Doink reminds Marc of his dog Jethro, who was also clever at snatching food from under the noses of his friends, particularly his canine friend Sasha. While Jethro knew that Sasha was possessive of her food, he was careful not to rile her. He'd eye her carefully, watching for her to make the slightest move away from her bowl, then he'd quietly and quickly slink in, grab a few morsels, and gulp them down. After the theft, he would lick Sasha's muzzle, then stroll away as if nothing had happened. Sasha had no clue. Jethro was also quite adept at stealing Marc's food, a skill many dog owners no doubt find exceptionally well developed in their own companions. The similarities between pigs and dogs raises the question of how we can treat them so differently.

Leo the Lucky and Doink the Cantankerous offer personal examples that counter the misperception of pigs and other farm animals as stupid and unfeeling. In fact, the scientific literature on the cognitive and emotional lives of food animals is extensive and ever growing, like that for our canine companions. For example, a recent review of pig emotion cited well over a hundred papers on mood and emotion in pigs.[12] Like Leo and Doink, pigs display a range of personality types similar to humans, with some being extroverted and gregarious and others being more cautious about new friends and new experiences. Pigs experience prolonged mood states like depression and happiness and show empathy toward one another. Contrary to popular stereotype, pigs are naturally hygienic, carefully separating their sleeping quarters from the outhouse, and taking care to bathe and groom. In the wild, pigs can be highly social and are unlikely to display the hyperaggression often seen in intensive-farming systems. Pigs are also able to recognize other individuals, and use a wide range of olfactory and auditory signals to communicate mood and intention and negotiate social hierarchies. Studies of pig intelligence often conclude that the cognitive abilities of pigs are even more advanced than those of dogs. For example, Stanley Curtis of Penn State University trained pigs to play joystick-controlled video games. The pigs, he observed, are "capable of abstract representation" and can "hold an icon in the mind and remember it at a later date." He concluded, "There is much more going on in terms of thinking and observing by these pigs than we would ever have guessed."[13] As Johannesburg Zoo director and biologist Lyall Watson writes in *The Whole Hog*, "I know of no other animals [who] are more consistently curious, more willing to explore new experiences, more ready to meet the world with open mouthed enthusiasm. Pigs, I have discovered, are incurable optimists and get a big kick out of just being."[14]

Similarly rich literature exists for chickens, cows, sheep, turkeys, and a whole range of other animals used for food, and consistently shows these animals to have a highly evolved range of cognitive and emotional capacities at least equal to the animals we keep as pets and consider worthy of protection from cruel treatment.

So how can researchers know what emotions animals are experiencing? And more particularly, how can we know whether certain kinds of experiences, like constraint in a small cage or crate, elicit negative feelings? As we discussed in chapter 2, measuring physiological markers like heart rate or cortisol levels in an animal can provide evidence that animals are feeling heightened emotions. But these physiological markers only measure emotional intensity. And emotions can be intensely good (excitement or anticipation or lust) and intensely bad (terror, dread). Cognitive-bias tests can provide additional information, telling us something about the emotional valence, namely whether an emotion is strongly positive or strongly negative. As you'll recall from chapter 2, research in humans and animals has demonstrated that emotional states can influence cognitive processes, including memory, attention, and judgment bias. Depressed humans are more likely to judge ambiguous events negatively, as are people who are in pain. Cognitive-bias tests can also tell us some useful things about the well-being of farmed animals.

One example of how to infer mood state from cognitive bias in farmed animals is a study on pain and pessimism in dairy calves, specifically relating to disbudding surgery (having their horns removed). Heather Neave and colleagues trained calves to respond differentially to two screens of different colors, one red and one white.[15] The calves learned to touch the "positive" screen for a reward of milk and to avoid touching the "negative" screen, which action was punished by a "time out" with no milk. Ambiguous screen colors were then introduced randomly, alternating with the red and white screens. Researchers recorded how often the calves were willing to try the new colors, taking this as a measure of judgment bias. In other words, the more they respond to something new with curiosity instead of fear, the more positive an emotional state is indicated. The calves were then subjected to the operation and tested again with ambiguous screen colors at six hours postsurgery (the peak of pain behaviors and cortisol levels) and twenty-two hours postsurgery. In both tests, the calves showed a negative judgment bias, and were less likely to approach an ambiguous color than at presurgery.

It is worth emphasizing that the point of the study is not to assess whether hot-iron disbudding without anesthesia is painful, because that has already been established. Rather, the study tells us that a painful event has more than a momentary effect on calves' emotional state. In this case, the researchers only studied emotional changes over a twenty-two-hour period; negative emotional states could have persisted even longer.

If behavior is a window into how an animal is feeling, then the capacity of farmers or stock managers to assess how animals are faring will depend on careful observation. Much can be gained, in terms of animal well-being, if a human knows the animals and is closely interacting with them and observing their behavior. For example, the ears, nose, and eyes of a cow are an excellent window into how she is feeling. Research by Helen Proctor and Gemma Carder found ear movements can be a reliable, noninvasive measure of an individual's emotional state. The cow's ears project backwards in a more relaxed position after they are stroked, which the researchers took to be a positive and low-arousal emotional state.[16] Another study found that the white of the eye can be an indicator of emotion in dairy cattle. When cows are scared and frustrated, we see more eye white than when they are stroked and calm.[17] We also know that a cow's nose can tell us about their emotional state. There is a decrease in nasal temperature as their stress level falls.[18] To pick up on these behavioral cues, however, the human caretakers have to know their animals and be paying attention. When Marc had the great fortune of meeting Bessie, a rescued dairy cow at Farm Sanctuary in Orland, California, he didn't look at her ears, nose, or eyes, but he knew immediately that she savored his companionship. When he sat down next to Bessie, she leaned her big head into his shoulder as he told her how beautiful she was.

As farming methods become more intensive they also tend to become more automated and efficient, with human workers having less contact with the animals. One might be tempted to consider this a welfare improvement, since so many of the interactions that animals in food production have with humans are unpleasant and involve yelling and hitting and electric prods and pain. Indeed, the negative impacts of human-animal interactions on farm animals has been long established; we know that rough handling and unkindness causes animals to feel fearful and stressed.[19] Yet animals can also benefit from increased interaction with humans, as long as this interaction is of the right sort, a point we'll return to later in the chapter.

Productivity as Proxy

A great many welfare studies are aimed at understanding what animals want and need, both in terms of physical care and, more recently and importantly, in order to keep them psychologically happy, or at least mentally sane. But often this research isn't only or even primarily about welfare, but is instead focused on maximizing productivity. When welfare science bleeds into industry practices, money rules and animals suffer.

Ignoring a wealth of scientific data to the contrary, animal scientists sometimes claim that we cannot measure emotions, and they go on to say that we can only manage what we can measure. Therefore, they claim, we should focus on "performance indicators" as an objective way of determining how animals are feeling, the idea being that poor welfare leads to poor performance (for example, poor growth, poor egg output, poor meat quality). Productivity becomes a proxy for welfare. The underlying assumption is that animals are in a poor state of welfare only when their physiological systems are so disturbed that reproduction and survival are impaired.

Oklahoma State agricultural economists F. Bailey Norwood and Jayson Lusk, among others, challenge the "productivity as proxy" logic. "If animal welfare could be inferred directly from the productivity and profitability of the farm," they write, "the egg dilemma would be no dilemma at all because birds are healthier and more productive in a cage system."[20] Hens in cage systems produce up to 270 eggs a year; cage-free hens only 259 eggs per year. But welfare in a cage system is impaired more severely than in the cage-free system, so why are the happier hens actually producing fewer eggs? One reason is that cage-free hens are busy engaging in more natural behaviors and are laying eggs on a less strenuously induced schedule. Also, hens in a cage-free system are allowed to walk around, and the increased exercise diverts some of their food energy away from laying eggs. Norwood and Lusk go on to say, "Productivity is a useful indicator of welfare, but only when applied to a *single* animal, rather than a group of animals or an entire farm."[21] Farmers thus may have to trade individual animal welfare for overall group performance. "When farms are constrained by land and building size, and machinery, it often makes economic sense to use more less-happy animals than fewer happy animals."[22]

Another drawback of the "productivity as welfare proxy" logic is that small improvements to welfare may increase productivity, leading to what seems like a win-win situation. The problem is that the "win" for the animals is pretty damn trifling. For example, hens who are given slightly larger cages become more productive: increasing the space, per hen, from forty-eight to sixty-seven square inches leads to more eggs. A chicken enclosed for life in a sixty-seven-square-inch wire cage cannot be said to have anything close to a decent life. Yet it is a convenient myth for industry to promote: look how productive our animals are; they must be really happy. As Ruth Harrison warned in *Animal Machines*, "For the factory farmer and the agri-industrial world behind him, cruelty is acknowledged only where profitability ceases."[23]

The uneasy tension between welfare and productivity is on full display in the milk industry. In what might seem like a victory for animalkind, there are fewer milk cows in production now than there were in the 1980s—almost two million fewer according to a *Washington Post* report. And yet dairy farmers have been able to keep milk-production levels stable. What this means is that more milk is being squeezed out of each cow, through a combination of design tweaks such as what cows are fed and when they are milked and aggressive genetic manipulation of the cows' anatomy. These milk-heavy dairy cows are bigger than "normal" cows, have higher and larger udders, and even have differently shaped legs. They also suffer from a slew of "adverse health outcomes." The welfare of these milk machines can be so compromised that even Temple Grandin has spoken out against what she calls "the bad dairies." She notes, "They make up most of the farms in the United States, and their cows are so wrecked by the time they stop milking they can barely be used for beef."[24]

Meat Science: Scientizing Cruelty

Welfare science both assumes and reinforces a particular moral and political point of view, namely that animals are here for us to use as we wish and that imposing significant harms on animals is acceptable, as long as humans are benefiting. Cruelty, in the context of food animals, is defined by the industry itself, and when the public expresses doubts about certain practices, the industry comeback is science: We've studied it and this is the best way to do things. You have to put sentimental feelings aside.

Meat Science magazine is a case in point of welfare science and industry being blended into one. The purpose of this industry-and-welfare-science publication is stated on the journal's main webpage:

> The qualities of meat—its composition, nutritional value, wholesomeness and consumer acceptability—are largely determined by the events and conditions encountered by the embryo, the live animal and the postmortem musculature. The control of these qualities, and their further enhancement, are thus dependent on a fuller understanding of the commodity at

all stages of its existence—from the initial conception, growth
and development of the organism to the time of slaughter and
to the ultimate processing, preparation, distribution, cooking
and consumption of its meat.[25]

Notice how completely "animal" is rendered into "commodity."
Here is the abstract of a representative research article from *Meat
Science*, on the pathophysiology of bolt stunning in alpacas. Yes, peo-
ple really study this sort of thing.

The aim of this study was to examine the behavioural and cra-
nial/spinal responses of alpacas culled by captive bolt shoot-
ing and the resulting pathophysiology of captive bolt injury.
Ninety-six alpacas were shot (103 shots) in a range of loca-
tions with a penetrating captive bolt gun (CBG). Ten (9.8%)
alpacas were incompletely concussed following the first shot.
No animals required more than two shots. Incorrectly placed
shots accounted for all of the animals that displayed signs of
sensibility. Damage to the thalamus, hypothalamus, midbrain,
medulla, cerebellum, parietal and occipital lobes were significantly associated with decreasing odds of incomplete concus-
sion. In conclusion, the study confirmed that CBG stunning
can induce insensibility in alpacas and suggests that the top
of the head (crown) position maximizes damage to structures
of the thalamus and brainstem.[26]

Within the context of meat science, this study wouldn't seem the
least bit out of place. Indeed, most of the articles in *Meat Science*
involve scientific measurement and quantification of practices that,
in any other context, would be seen as barbaric. But the fact that the
activity is scientized (this is meat *science*) allows it to maintain an
aura of moral neutrality.

Another *Meat Science* report analyzes the effects of chronic stress
on meat quality in ruminants (animals who acquire nutrients from
plants by fermenting food in specialized stomachs prior to digestion).
It is well established that chronic stress, an "inevitable consequence"

of the process of transferring animals from farm to slaughter, causes muscle glycogen depletion and "dark cutting condition" (that is, less-than-ideal meat quality). But what have been far less well studied are the effects of acute pre-slaughter stress on the quality of meat. "All meat animals," the authors note, "will experience some level of stress prior to slaughter and this, in turn, may have detrimental effects to meat quality."[27] The research is not about the effects of chronic and acute stress on ruminant well-being, but rather on meat quality. That causing harm to animals is acceptable is an unspoken (and probably unconscious) assumption. And the reasons given for remediating pre-slaughter stressors have nothing to do with the animals themselves. The "science" is all about the meat.

So, where is the line between welfare science and meat science? There really isn't any, and that's one of the most significant problems with the welfarist paradigm. Some welfarists are certainly motivated by compassion, but the compassion is intermixed with business and with promoting and rarely if ever questioning the agenda of the meat industry and the eating habits of consumers. And so the science will never and can never really be put into the service of animals, because the interests of animals most definitely conflict with the interests of meat producers and meat eaters, and most definitely take a backseat. Just as you can't take the cream out of the coffee, you can't take the industry out of welfare science.

Humane-Washing

This is as good a place as any to discuss the use of the word "humane," one of the most overused and meaningless in our current vocabulary. If you hear the word "humane," you can pretty well bet that something bad is happening to animals and somebody is trying to clean it up and make it look less ugly. Philosopher Joel Marks, in his "Animal User's Lexicon," notes the irony of the term, "since a cold-eyed comparison of the way human beings treat one another and the way other species—the 'beasts'—treat their own would hardly be flattering to ours." But even more important than its internal irony is the term's use by industry and welfare science alike, which "label all of the cruelty and death that are inflicted on the tens and hundreds of billions of nonhuman animals year after year as . . . humane!"[28]

Indeed, it does seem remarkable the frivolous way "humane" is thrown about as a term of endorsement for even the most unseemly intensive-farming practices. For example, in the United States the certification label ANIMAL WELFARE APPROVED allows three different methods of castration—surgical, rubber ring, and Burdizzo (which involves crushing the balls)—to be performed on cattle or pigs without anesthesia. According to the so-called Twenty-Eight-Hour Law, another humane protection, vehicles transporting animals (not including birds, who aren't legally protected) to slaughter must stop every twenty-eight hours to let the animals eat, drink, and move about. So allowing animals to stand in the bed of a moving truck, exposed to cold or heat and without water or food, is considered humane as long as it doesn't exceed twenty-eight hours at a stretch. The federal Humane Methods of Slaughter Act requires that "before being shackled, hoisted, thrown, cast, or cut, livestock animals must be rendered insensible to pain by being gassed, electrocuted, or shot in the head with a firearm or captive bolt stunner."

Perhaps the reason the above-cited slaughter methods can be called humane with a straight face is because they are defined by reference to practices that are even worse. One example of "worse" was highlighted by Nicholas Kristof in a *New York Times* editorial, which discussed the production process at one North Carolina chicken slaughterhouse. Because the factory disassembly line at this facility moved so rapidly, workers were often inaccurate with the knives that were supposed to slit the throats of birds hanging upside down by their feet, or they missed some birds entirely. The result was that a fair number of chickens were still very much alive when they reached the scalding tank. Now, *this* is inhumane—the live scalding of animals. But because humane-slaughter laws in the United States do not apply to poultry, the plant's operation was in line with industry standards. The live scalding of some of the birds was considered the price of an efficient production line.[29] So, "humane" is defined in comparison to worst-case scenarios. The birds that were killed "humanely" died swiftly, without being scalded alive. This happens on a small scale, as in the slaughterhouse example. And it happens on a larger scale, too.

It is unfortunate for the science of animal welfare that the language of "humane" is so frequently and persistently popping up in the literature, because it raises questions about the objectivity of the science. "Humane" is a dirty little lie. It's a feel-good word that means we're throwing animals a tiny bone of some sort, like a bigger cage or a curved chute leading to the killing floor. At one point, Grandin talks glowingly about a "relatively humane" kosher restraint pen (a pen that squeezes, lifts, and inverts an animal for throat slitting). She says, "If I visualize myself as an animal I would willingly walk into the modified . . . pens." We surely would not.

"But They Had a Life!"

The animal welfare literature is riddled with some absurd claims. One of the most egregious of these is that life in captivity is cushy and comfortable, whereas nature is harsh. Given a choice, animals would certainly pick being our domesticated servants, even if it means living in a concentrated animal feeding operation (CAFO). In itself, this is merely silly. But the claim, and others like it, has a tendency to get couched as science within the welfare literature, and this is where it becomes dangerous. We'll be identifying some of these "truisms" throughout the book, where they are most likely to pop up (e.g., "death is not a harm" with respect to laboratory animals, and "life in captivity is safe and easy" in regard to zoo and companion animals).

One common line of argument is that we have done food animals a favor by "letting" them be born. After all, we gave them a life. And this life, if we visualize the cow on the side of the milk bottle or the pig in the advertisement for bacon, is lived with grass underfoot and sunny skies overhead. Our go-to farm-animal-welfare icon Temple Grandin engages in this godlike fantasy. "One day," she writes in her manual *Humane Livestock Handling*, "I was standing on a long overhead catwalk at a stockyard and chute system I had designed.

As I looked out over a sea of cattle below me, I had the following thought: These animals would never have been born if people had not bred them. They would not have known life."[30] She says, in other words, that the cows owe their existence to the fact that people want to slaughter and eat them.

We can leave it to the philosophers to parse the logic and the ethics of this claim. For our purposes, it is enough to simply remind the reader that although Grandin is a scientist, this is not a scientific statement. Neither is this gem, also from Dr. Grandin: "Death in a well-managed, well-designed slaughterhouse is much less frightening or painful than death will likely be in the wild."[31] She doesn't provide any science to back up this statement, probably because she can't. This is pure, unbridled speculation. Very little research has been done on how animals die in the wild, so we can't really quantify how frightening or painful it might be for them, nor can we provide any meaningful comparisons.

Freedom Enhancement in the Food-Animal Setting

The Five Freedoms acknowledge that behavioral freedom—the freedom to "express normal behavior"—is important for animals. But the principle of respect for behavioral freedom must be taken considerably farther than the welfarist paradigm suggests. Not only should chickens be able to flap their wings (a normal behavior), they should have the freedom to go about their daily lives; they should be able to engage in social interactions, make choices about what to do, who to lie with, and when to eat and poop.

Although food animals will never be entirely free, there are ways in which we could make meaningful improvements that would enhance the range and quality of freedoms they can experience. Even now, some farmed animals have greater freedom than others. On smaller-scale farms, animals are often allowed to forage relatively freely, are allowed much greater freedom of physical movement, and have a much greater measure of control over their social interactions.[32] Even some large farms seek ways to improve real freedoms.

An interesting example of trying to pay attention to behavioral freedom is found in an essay written by British researcher Kate Millar. Millar's title seems to promise much: "Respect for Animal Autonomy in Bioethical Analysis." The subtitle is "The Case of Automated Milking Systems (AMS)."[33] Automated or robotic milking systems allow cows to milk "voluntarily," at least in principle. In an automated system, cows enter a milking unit on their own initiative. An identification sensor reads a tag attached to the cow. If she has milked too recently, she will be ushered out of the milking system. If she is a candidate for milking, robotic arms will clean her teats, attach suction cups, and clean the teats again after milk has been collected. There are several different systems, including free choice, rewarded access, and obligatory access, with rewarded access being the most common. Of course, these systems are not designed to give cows autonomy so much as to extract milk without human labor and increase profit.

Whether or not an AMS is really voluntary is open for debate. The desire to be milked does not generally provide sufficient motivation to the cows, so some kind of reward such as extra rich grain at the milking station must be used. And there are potential welfare trade-offs. For example, when an AMS is in place, cow welfare is no longer monitored by humans who spend hours with the animals during milking, but is instead monitored by data collected through the milking machine (such as on quality of milk and body temperature). And of course, an AMS offers cows only a small improvement in terms of their relative freedom. But a small improvement can make a difference with respect to animal well-being.

It is important to note that some seeming freedom enhancements, like "cage-free" eggs from "free-ranging" chickens, are just humane-washing of an ugly truth. "Cage-free," for example, often simply means that the chickens are not in battery cages but are "free" in a large enclosed industrial barn. They may still have an average of only one square foot of space each. A "free-range" chicken has hypothetical access to the outside, which may be little more than an open concrete "yard." The access is only hypothetical because a very small number of birds actually succeed in getting out.[34] Contrast this with the chickens who really are free ranging, like those who live with the Hobbses at Happy Mama Acre. These chickens are never confined, live in small flocks which they form based on their own social preferences, and can feel the earth under their feet and the sun on their feathers.

One of the crucial "small improvements" that could be made would be the elimination of extreme forms of confinement such as gestation crates for pigs. Breeding sows are housed in crates scarcely larger than their own bodies, so they are unable to turn about, a clear violation of the third freedom. They must eat, sleep, and defecate in this tiny space. The concrete floors damage bones and feet and joints. Psychological deprivation is profound: known to be as intelligent as dogs, pigs are bored out of their minds and frustrated in these crates, and will bite and chew at the bars and rub themselves bloody. As a Humane Society of the United States report notes, "The amount of time sows engage in stereotypies increases with time spent in crates. This expression of abnormal behavior is widely accepted as a sign of psychological disturbance. By comparison, in situations where sows have greater freedom in more complex environments, the amount of stereotyped behavior is nearly zero."[35] As of this writing, nine US states and Canada have banned the use of gestation crates for sows. This is a huge step in the right direction.

Well-Being Means Paying Attention to Individuals

One of the most important elements in the science of animal well-being is that we begin and end with a focus on the individual. Welfare science tends to fall short in this regard, as too often does the broader science of animal cognition and emotion. Scientists treat animals as homogeneous groups. We study what pigs are like, or what sheep are like, or what cows are like. Yet we know from our human experience that needs and wants can vary quite a bit from one person to another. The same is true for individuals of any other given species of animal. Each individual is unique, and has particular likes and dislikes, and is shaped by his or her own experiences and personality. We must be very cautious when talking about "the cow" or "the pig" or "the chicken."

In preference studies there is a strong tendency toward this kind of averaging of behavioral responses that masks individual differences. But this is scientifically dicey, because what you get is simply an averaging of behavior over the group. Even when studies are conducted on very small groups of animals, which often they are (a scientific problem in itself), the published results offer a generic "pigs like this and not that." In a study of ice cream preferences of six pigs, where the choices are severely limited (as they often will be in welfare science), we get an average of what six different pigs want, given a choice between two options. If two pigs strongly prefer beet while four strongly prefer sweet potato, the published result will be: "pigs prefer sweet potato ice cream." Thereafter, "pigs should have sweet potato" will be the welfare rule of thumb. Woe to the beet lovers. A study of "animal-material interactions" in 128 pigs, to take a real-life example, concluded that pigs like playing with a suspended rope better than chain.[36] If we were to generalize about what human beings are like from a study of 128 individuals, scientific eyebrows would certainly be raised. Yet studies of nonhuman animals generalize like this all the time. Often, the number of "subjects" being studied is a tiny handful of animals. And welfare guidelines and practices are often based on these crude aggregate statistics, which then fail to offer consistent welfare improvements. Those animals who fall into the zone of average may feel some benefit; but outliers will not.

One way to counter this is to make sure that studies of animal behavior, and particularly, for our purposes, studies of preferences and aversions, pay careful attention to *individual* differences. Taking the time to observe a group of six pigs, as they choose from a series of different types of food or toy or social interaction, could be exceedingly useful, as long as the researchers were trying to understand the pigs as six unique individuals and not trying to draw conclusions about pigs in general. Studies trying to understand species preferences and aversions, or the "species-typical behavior" the Five Freedoms idealize, should never try to deliver more punch than the study size and resulting data support, which is often pretty minimal. Moreover, we should be wary of making generalizations about species-typical behaviors and preferences, when attention to nuance is essential to animal well-being.

Another reason to pay attention to individuals is that there is a certain moral power that comes from rallying around an identifiable animal "person" rather than an abstract mass of animals. *New York Times* columnist Nicholas Kristof makes just this point in an essay called "Animal Cruelty or the Price of Dinner?" Kristof recounts how people got incredibly upset when a Florida man dangled his dog by the neck over a second-floor balcony and threatened to drop the dog twelve feet to the ground. Onlookers tried to intervene, and someone in the crowd caught the incident on video and posted it on Facebook. The clip went viral and incited public outrage. The local police responded promptly and arrested the man on charges of animal cruelty. Kristof ends his piece with a challenging question: "If we can rally on behalf of a frightened dog in Orlando, can't we also muster concern for billions of farm animals?"[37] This harkens back to the quote with which we opened this chapter.

Naming Matters

Ironically, stock managers seem to know, perhaps better than scientists, that getting familiar with individual animals is important. On some farms, animals are known not by number, but by name. Some might claim that this naming of animals is bad business, because animal husbandry involves, at some point, putting aside your feelings for an animal and doing what needs to be done. But perhaps surprisingly, these farms often function better than those where animals are faceless units of production.

A study by Catherine Bertenshaw and Peter Rowlinson explored the perceptions of stock managers toward the human-animal relationship on dairy farms.[38] The predominant relationship between humans and cows is one of fear. The cows are afraid of the humans. This fear-response affects the productivity and welfare of the cows. In their survey, Bertenshaw and Rowlinson found that the vast majority of farmers consider cows to have feelings and to be intelligent. On farms where the stock manager knew every animal as an individual and gave each cow a name, milk yield was 258 liters higher than on farms where cows were not individualized.

Smaller-scale farming has the potential to be much better for the animals because farmers—or stock managers, as perhaps they must be called—are more likely to be able to attend to individual animals. This is impossible within the massive intensive agricultural facilities which are run day to day by unskilled and low-paid workers who don't know the animals, and who have to be brutal in order to process product fast enough to meet company expectations, and are thus unable to acknowledge the inherent worth and dignity of the animals.

Welfare research is increasingly paying attention to the quality of human-animal interactions, and the findings generally confirm that treating animals well is better for everyone involved. A 2015 study found that calves who were gently stroked by people early in their life gained weight faster, were less fearful of humans, and had lower heart rates—and they seemed to enjoy the attention![39] Farmers using "low-stress weaning" of calves, where mother and child are allowed nose-to-nose physical contact through a fence-line during the weaning process, report that animals experience less distress (less bellowing) and that calves gain weight faster.[40] These improvements could be viewed cynically as improving productivity by leading to faster growth. But there is no question that they also offer small improvements in the lives of these animals.

Eating Comes First

Philosopher Joshua Knobe, interviewed in *Salon*, remarks on a common perception embodied in a great deal of writing about animal ethics: we treat animals the way we do because we have certain attitudes toward them. The opposite is true, he argues. "We think of animals the way we do because we eat them, use them . . . The fact that people eat meat gives them certain philosophical ideas, it's not that philosophical ideas make them eat meat."[41] Knobe is partly right and partly wrong. He is right that eating comes first; we adopt a certain way of eating and make our philosophical ideas about animals match up with what we want on our plates. Indeed, a study reported in *Personality and Social Psychology Bulletin* demonstrates that people tend to engage in "mind denial" in relation to the animals they plan to eat. Animals destined for people's plates are ascribed diminished mental capacities, and this mind denial "reduces negative affect associated with dissonance."[42] To take Knobe's suggestion one step further, we make our decision about what or whom to eat and then shape welfare science such that it supports this choice.

But he suggests that the process works in only one direction: eating first, then ethics (and science). But certainly the process is more circular. Most children are raised as carnivores, and only later do some opt out of eating meat or using animals. So eating habits are passed down along with the values that support them, but a good many of us come to question both our meal plans and the philosophy behind our choices. Knobe may be right about the direction in which the circle turns, but we like to believe that there is room for evolution and change.

Why Good Welfare Will Never Be Good Enough

Grandin's stairway to heaven is a potent symbol of why good welfare cannot and never will be good enough. The stairway might be a minor welfare-enhancer, but it is not a freedom enhancer. Some of the animals being raised for their flesh may suffer some small percentage less pain and terror and may, in the best of welfarist worlds, even enjoy fleeting moments of joy. But welfarist reforms like the stairway don't move animals one inch closer to being free from exploitation. In fact, the stairway and similar welfarist reforms may actually make the plight of animals worse, because they ameliorate any negative feelings people may have about eating steak, for example, and because they lend support to the idea that science is making animals safe. Welfarism is not enough, and nowhere does welfarism fall as short as in relation to the billions of food animals.

The European Union Welfare Quality report recommends thirty to fifty measurements of welfare for each species of farm animal.[43]

This is appropriate, because animal welfare is complex and there are many factors to take into consideration. The idea seems to be that the more measures we make, the more carefully we scientize them, the better off animals will be. This is true, to an extent, of course. But it has to be seen within the broader picture, which is that we impose significant harms on these animals. Welfare has to do with how well an animal is coping within the environment we've placed her, such as how well the battery-caged hen is enduring the profound assaults of confinement. The fact that an animal has to cope suggests that these environments are challenging, unnatural, ill-fitting, and cruel.

As valuable as improvements in welfare are, the conversation about respecting animal preferences needs to dig much deeper than it currently does. And the results of these discussions must translate into significantly better care for the animals being "translated" into food on our plates. We need a major transformation in cultural attitudes toward food animals.

As we noted before, scientists are learning a great deal about the cognitive and emotional lives of farm animals. But this knowledge is not being translated into any kind of evolution in our moral attitudes toward these creatures, although it seems inevitable that it should. How could the knowledge that baby chicks can count and that pigs feel empathy and that cows have maternal bonds with their infants every bit as strong as we do not make people increasingly uncomfortable about the way animals are treated within intensive-farming systems? People often acknowledge it, but then go on consuming these sentient beings as if they didn't know better.

We need a science that is shaped by carefully articulated and explicit values, especially the value of individual animal lives. We also need to pursue reforms that are freedom enhancing for animals, and we need to work within a paradigm of science that truly seeks independence from industry and profit. We can seek to increasingly align profit with welfare, as Wayne Pacelle, Humane Society of the United States president and CEO, argues in his 2016 book *The Humane Economy*. Even better though, would be to align profit with individual well-being and to put an end to industries that profit by abusing other animals.

As studies of animal cognition and emotions grow ever more mature, the scientific paradigm will need to shift to accommodate new perspectives on who individual animals are and what they want and need. The welfarist paradigm will continue to crack and show signs of strain, and the paradigm of well-being and freedom will have room to develop and flourish.

CHAPTER 4

Fat Rats and Lab Cats

Although the Brambell Report initiated the development of animal welfare science into a distinct field, the first seeds of welfare science had been planted a number of years earlier. Richard Haynes, in his history of the animal welfare movement, traces its beginnings to the Universities Federation of Animal Welfare in London in 1926, and founder C. W. Hume's belief that "animal problems must be tackled on a scientific basis, with a maximum of sympathy but a minimum of sentimentality."[1] In 1946 UFAW published what was likely the first handbook on the welfare and management of experimental animals.

The Three Rs

The Brambell Report was a key catalyst for the scientific study of farm-animal welfare; for laboratory-animal welfare, the primary mover was William Russell and Rex Burch's 1959 *The Principles of Humane Experimental Technique*. Like the Brambell Report, Russell and Burch's work was revolutionary for its time because it assumed, as did Brambell, that animals have feelings and that these feelings matter. The notion that animals have no feelings is "pathological" thinking by humans, the authors attest, and "an unattractive perversion of the Golden Rule."[2] They go on to say, "We shall not waste any time on those philosophers who would forbid us to speak of consciousness in nonhuman animals."[3] Humans are, by their nature, empathic toward animals, and even scientists engaging in experimentation on animals care about their subjects.

Just as the Brambell Report was concerned with serious welfare issues raised by the increasingly intensive and industrialized methods of food production, Russell and Burch's report raised concerns about what was happening within an increasingly intensive and industrialized research juggernaut. They noted that the number of animals used each year in research had been steadily rising since at least the 1940s, and worried that "biology is by now an industry." Moreover, it is an industry that has not always been good about "corrective feedback" or "systematic scanning of its techniques." The result is a good deal of inefficiency and waste, as well as poor science.[4] Despite the generally good intentions of scientists within the current scientific and moral paradigm of the time, a number of procedures being carried out on animals were, according to Russell and Burch, simply inhumane. In addition to appealing to the empathy of scientists, they warned scientists that the "wages of inhumanity are . . . paid in ambiguous or otherwise unsatisfactory experimental results."[5] In 1959 this was a groundbreaking claim.

Since Russell and Burch's work preceded Brambell by six years, there was no mention of the Five Freedoms.[6] Instead they articulated what has become a kind of mantra for laboratory-animal welfare: the three Rs: reduction, replacement, and refinement. Reduction focuses on quantitative aspects of pain and fear: we should look at both the extreme unpleasantness of an experiment and the sheer number of animals, or a combination of the two, and seek to reduce both. Replacement encourages researchers to replace sentient animals with nonsentient forms of life such as cell cultures, to replace "higher" animals with "lower" (e.g., swapping primates for mice or nematodes), or to use nonanimal alternatives such as computer models, plasticene models, cell cultures, and videos. Refinement involves sometimes subtle alterations to experimental design, such that suffering and distress are minimized (e.g., setting clear end points, at which pain is considered too severe and an animal must be killed, or using anesthesia during surgical procedures).

As we saw in the previous chapters, welfare science is guided as much by value assumptions as by science. According to welfarism, what's good for animals is a matter of scientific investigation, and scientists are the best judges of what animals need. This means that the value assumptions of science, namely that animals are here for us to exploit and that imposing harm is a scientific necessity, go unquestioned. Haynes says that "disciplinary assumptions frequently have had the effect of prescribing what the public ought to find acceptable."[7] And indeed, just as industry representatives like Temple Grandin have defined welfare for farm animals, so too have laboratory scientists and those representing industry such as pharmaceutical companies controlled the dialogue about the proper use of animals in the experimental setting, deflecting attention from the broader value assumptions that reinforce the status quo. Perhaps the most striking example of this is that the industry response to any moral hesitation about animal research is "What if your daughter had a dread disease, would you sacrifice her life to save a lab rat?" This sets up a false dichotomy, suggesting to a concerned public that the "sacrifice" of animals is essential to medical progress and that we have to choose between children and animals. The tyranny of this industry perspective is also evident in the fact that mice and rats, who are used by the billions within research, are still unprotected by the US federal Animal Welfare Act.

Unlike the meat and dairy industry, where there is an attempt to create an illusion that the animals on their way to market or on their way up the stairway to heaven are happy, or even grateful to us for giving them a life, this line of reasoning is blessedly absent from discussions of laboratory animals. There is certainly a good deal of humane-washing and massaging of public information to play down or hide the horrors that animals experience and play up the enormous benefits to humans. But even the cagiest animal researcher will not deny that animals are often harmed in serious ways by the research enterprise. Endemic among research scientists, however, is a certain Claude Bernard–esque blindness to the "blood that flows."

Dumb and Dumber: Does Suffering
Correlate with Intelligence?

Why does the use of "higher" species like primates and dogs and cats garner so much public attention, while the use of "lower" animals like mice and rats does not? The answer is, most people recognize that the suffering of animals who are most like us is significant. This is one reason that the first of the three Rs is to replace higher species with lower, with creatures who seemingly have less to lose. For example, the Institute of Medicine does not simply recommend that primates be phased out; they advocate that mice be used instead. And indeed, one of the most notable trends in animal research, at least in the United States, is a rapid and significant spike in the number of rodent research subjects. On the one hand, the replacement strategy has been informed by science: the more we know about the capacities of primates, the more serious the ethical implications are of using them in biomedical research. On the other hand, the replace strategy is also surprisingly unscientific, because it assumes that intelligence can usefully be ranked from low to high and that greater intelligence is correlated with greater suffering.

Replacement assumes a hierarchy of "higher" to "lower" species, with higher having more significant interests. Despite the intuitive appeal, this is morally and scientifically complicated. For example, the replacement of "higher" apes with "lower" rats is based, at least in part, on judgments about the intelligence of the two animals. Apes are like us, so obviously they must be supersmart. And rats, well . . . they aren't even "animals." But, as any good animal-behavior scientist will tell you, all animals are intelligent in their own way, and cross-species comparisons of intelligence are fraught with errors. As Hungarian anatomist János Szentágothai famously remarked, "There are no 'unintelligent' animals; only careless observations and poorly designed experiments."[8]

Furthermore, regardless of how intelligent individuals of a species are taken to be, intelligence itself is not what really matters. What matters, as philosopher Jeremy Bentham famously said over two centuries ago, is whether or not an animal can suffer. The idea that great apes and dogs suffer more than rats and mice is based on an assumption that brain size is directly related to suffering; that is, individuals with larger brains suffer more than those with smaller brains. But cross-species comparisons of suffering are just as fraught as comparisons of intelligence.[9] It is very difficult to quantify how much an individual animal or human actually suffers, and judgments about whose suffering matters most are arbitrary, not to mention bizarre. Is there a superior kind of suffering? Doesn't the suffering of each individual matter?

The logic of the replace strategy is good: we should pursue research that will impose the fewest possible harms, by replacing intelligent animals with nonanimal alternatives. We just need to be careful with our judgments about who is intelligent and who should replace whom. Ultimately, we need to move beyond unscientific terminology like "higher" and "lower." It may be that the animals with whom people most strongly identify, whether because they are very much like us or because they are our loyal companions, are those who will earn protection first, and we should celebrate these victories. But we cannot and must not stop there. Ideally, we can hope for a time when every "who" will be replaced with a "what." Even mice can look forward to a day when they will be replaced with a digital organ-on-a-chip that simulates the physiology and mechanics of an organ or organ system.

Why Science Isn't Helping Research Animals

It is surprising that the large amount of available data from detailed research into the cognitive and emotional capacities of animals has not significantly influenced laboratory-animal welfare. One exception, which we will discuss, is that some scientists have paid careful attention to the effects of poor welfare on the quality of scientific data, and this has led indirectly to welfare improvements. Otherwise, the knowledge-translation gap is surprisingly wide. Regulatory protections for lab animals lag far behind what we know, and moral attitudes toward animal research have been resistant to influence from the science of animal cognition and emotions.

Consider, for example, the situation of rats. Rats are one of the most common model animals in research, particularly for the study of mood disorders such as anxiety, depression, fear, and hopelessness. We have an encyclopedic knowledge of rat emotions.[10] We know they feel pain and that they experience many of the same negative emotions we do. Research suggests, for example, that rats housed in isolation may develop depression.[11] We also know that rats experience positive emotions such as pleasure. Neuroscientist Jaak Panksepp has spent decades studying emotional states in rats and examining thin slices of brain tissue for neurochemical changes. He has found, for example, that rats "laugh" when they are tickled by human handlers and when they engage in play with cage mates.[12]

Japanese researchers have offered further insight into rat emotions by studying helping behavior. The researchers put one rat into a tank of water from which there was no escape, one of the common rat models used to create a state of panic. Another rat was placed on a platform, in view of the drowning rat, and had to figure out how to push open a door that would help the waterlogged rat escape to dry ground. The rats were highly motivated to open the helping door even when offered the choice of earning chocolate instead.[13] These are interesting results, but knowing from the get-go that rats experience panic, why would you subject them to the forced swim in the first place? The findings of Panksepp and others who have studied rat emotion and behavior have made little difference in researchers' willingness to employ rats in invasive research, nor have they influenced welfare protections for rats.

Despite all that we know about the intelligence and emotionality of rats, they have none of the measly protections afforded to the sentient creatures who made the Animal Welfare Act's list of "animals." The AWA is written explicitly to "exclude birds, rats of the genus *Rattus*, and mice of the genus *Mus*, bred for use in research."[14] Clearly the AWA is not a reflection of modern science, since even first graders know that rats and mice are animals. Legal protections for animals are like molasses—very slow moving. It took about fifty years of "data collection" for the AWA to acknowledge that psychological well-being is as important to animals as physical well-being, and in 1985 AWA required laboratories to provide for the psychological well-being of primates. This requirement has yet to expand to nonprimate species.

The situation of rats and other noncharismatic species is not helped by the fact that they garner little public sympathy, and because few people have a chance to get to know them personally. Many people, when they think of rats, imagine them slinking around urban dark alleys and sewers or, alternatively, in their "natural habitat": a spick-and-span silver cage in a shiny research lab. Although we aren't advocates of pet keeping, it would likely help the situation of rodents in research if more people kept rats as pets and had a chance to interact with them up close. Jessica's daughter Sage was a great fan of rats, and over the course of her elementary school years had a sizable collection of pet rats. These rats served as ambassadors of a sort among the family's neighbors and friends. Other children, who hadn't yet absorbed the prejudices of their parents, were eager to meet the rats and were delighted by their curiosity, friendliness, and smarts. Parents almost always reacted with disgust, and with a "Ew! Their tails are so disgusting!" But those who spent even a little time getting to know Velveeta, Henry, Midas, and Fuzzies were quickly won over.

Problem: Stereotypies / Response: Enrichments

The small white mouse in a shiny silver cage is the quintessential research animal, and for good reason. Rodents, and particularly mice, have long been science's most popular living and breathing petri dish, and their popularity continues to grow. They are used in every corner of biomedical research, from the study of cancer mutations, to brain function, to drug testing. Unlike nonhuman primates, dogs, and cats, where confinement to a small cage can appear to the public as a serious welfare concern, caged mice tend not to garner the same negative reaction. But mice are not machines; they are living, breathing, feeling burrowing creatures.

Mice in laboratory cages are not necessarily happy. At least half of the mice in standard laboratory cages develop stereotypies.[15] Dr.

Georgia Mason at Canada's University of Guelph, one of the world's experts on stereotypies, offers this definition: "Stereotypic behaviors are repetitive behaviors induced by frustration, repeated attempts to cope, and/or CNS [central nervous system] dysfunction."[16] We wrote about stereotypies in chapter 3, but include Mason's definition here to emphasize that stereotypies reflect damage to the central nervous system, caused by the psychological rigors of confinement. Let's say this clearly: confinement causes pathological changes to the brain.[17]

Stereotypies develop over time, particularly in young animals, and must be understood as causally related to captivity, as they are completely absent in the wild. "Stereotypies," says Hanno Würbel, "generally seem to develop from adaptive behavioural responses aimed at coping with the chronic thwarting of highly motivated behaviour."[18] Behavior patterns that are crucial for survival in the wild, such as seeking shelter, exploring olfactory cues, returning to mother after weaning, and species-typical motor behaviors such as digging, hiding, foraging, and caching, are continually thwarted in captive environments. The thwarting of responses to environmental cues "triggers sustained appetitive behaviours and/or intention movements," such as jumping, running, bar mouthing. "The barren nature of the environment may primarily act through the impairment of normal brain development, resulting in impoverished brain architecture which may predispose animals to stereotypy development."[19]

Confinement in a cage is a serious welfare concern with research animals, and one for which there is no good fix. As Viktor and Annie Reinhardt note in their book about the welfare of rodents and rabbits, "Being confined in a barren primary enclosure is probably the most serious stressor for animals."[20] These animals have no environment to manipulate. A barren cage deprives mice of an entire range of highly motivated behaviors such as nesting, hiding, exploring, and searching for food. Confinement is associated with high levels of anxiety and fear, behavioral pathologies such as hair pulling, cage biting, and sham digging, alterations in brain neurochemistry, and hyperaggression. Space has little value to an animal, they note, unless it is shared with others of their kind and unless it is furnished with structures and objects that are meaningful to the particular species. Stereotypies and other maladaptive behaviors shouldn't be called "abnormal," they argue, since "it is the species-inappropriate conditions under which the subject is forced to exist that are really abnormal, not the subject's attempts to adapt to them."[21]

Recognizing some of the causal factors for development of stereotypies, scientists have tried to respond by developing enrichment strategies. Ron Swaisgood and David Shepherdson, for example, offer the following problem/response list:

1. Problem: frustrated motivations to perform specific behaviors;
   Response: cages or enclosures that mimic nature; meeting specific frustrated motivations (e.g., targeting foraging behaviors, making animals hunt for their food).
2. Problem: paucity of behavioral opportunities;
   Response: increasing the physical complexity of the environment.
3. Problem: lack of sensory stimulation;
   Response: increasing sensory stimulation (e.g., odors, noises, visual complexity, food textures).
4. Problem: stress;
   Response: removing sources of stress or providing coping options (e.g., providing hiding places), and providing enrichments that give the animal control.[22]

These are all examples of what has become known in the welfare literature as environmental enrichment. Enrichment is defined by Vera Baumans as "any modification in the environment of the captive animals that seeks to enhance their physical and psychological well-being by providing stimuli meeting the animals' species-specific needs."[23] The National Research Council defines enrichment as "a method to enhance animal well-being by providing animals with sensory and motor stimulation, through structures and resources that facilitate the expression of species-typical behaviours and promote psychological well-being through physical exercise, manipulative activities and cognitive challenges according to species-specific characteristics."[24] The goals of enrichment, then, are to enhance physical and psychological well-being, increase the expression of some behaviors (species-typical behaviors), and reduce expression of other behaviors (e.g., stereotypies).

Yet as we learn from the 868-page text *The Laboratory Mouse*, it is difficult to know exactly what kinds of things lab mice want and need. Generations of targeted breeding of lab mice may have changed their biology. So what do we use as the benchmark of mouse well-being? It isn't clear. In transgenic mice, diversity of traits bred for leads to diversity in behavior, physiology, and response to the environment. Environmental conditions that may be optimal for one strain of mouse aren't for another. Maybe this is one reason the enrichment literature for the laboratory mouse is so full of self-serving and contradictory findings.

Even optimal cage size is elusive because there are so many variables that might be measured and that might make a difference (type of material, height, shape, volume, visual effects of caging material, acoustic effects of caging material, lighting, ventilation, and so on). Cage or shelter color even makes a difference. Mice are dichromats, which means they have two types of color receptors and perceive color in the yellow-green and ultraviolet ranges. Because of this, they tend to like white cages best and red cages least. Nest boxes are valuable, but the material from which they are made makes a difference in mouse preferences. Even the types of toys or entertainments can be surprisingly important. Introducing a wheel changes the time budget for mice although it does not reduce stereotypic behavior. Adding wheel running can alter behavior and physiology, and thus alter results of scientific studies.

In an article on the neurobiology of stereotypy, University of Florida psychiatry professor Mark Lewis and coauthors warn that "despite the popular use of the term environmental enrichment, attempts to make captive environments more complex hardly result in environments that could be considered enriched relative to the animal's natural habitat."[25] Nonetheless, they admit that even small enrichments can have welfare benefits and reduce stereotypies. Deer mice reared in larger and more complex housing develop fewer stereotypies than standard-housed mice. Studies on rats, mice, bank voles, and various other species have shown that being raised in more complex environments has significant effects on brain development. And enrichment has a variety of other central nervous system effects such as improved spatial memory, improved retention of learned tasks, and improved memory in fear conditioning. Research has also shown decreased voluntary alcohol consumption and decreased aggression in animals in enriched environments.[26] Indeed, we've known about the effects of enrichment on brain development in animals since the 1940s, when Professor Donald Hebb raised a group of rats in his home, in a complex cage environment that was like a great big Disney hotel for rats. The home-raised rats were better at learning mazes and had larger brains than their lab-raised counterparts.

Environmental enrichment is one way to improve the welfare of animals. Another perhaps equally important enrichment is social. Indeed, it is not just barren cages that create the stresses that can develop into stereotypies. Animals can be subjected to complete social deprivation when reared and housed in isolation. They can also be stressed by social separation when they have been cohoused and allowed to develop social relationships, but are then abruptly removed from their companions. Furthermore, they can be stressed by separation of mother from offspring. Thus, the lone mouse in a metal cage has been replaced, in a great many research settings, with groups of mice housed together.

The Harm of Death

In chapter 3 we suggested that there are certain truisms built into the welfarist paradigm that persist and are replicated over and over, like nasty viruses. One of these is that death doesn't really harm animals. This belief comes into play in all areas in which we interact with animals, and could be mentioned in every single chapter. But in the realm of laboratory animal welfare science, the notion that "death is not harmful" is particularly interesting and complex. In some respects, the death of animals is taken far more seriously within the research context than in many other realms of animal use—which may seem strange, because billions and billions of animals are deliberately being killed. In many research protocols, the manner and timing of death is a crucial aspect of experimental design, and the collection of meaningful data will depend on carefully orchestrating the "end points" of the study.

Still, as a welfare concern, death is treated as if it were relatively benign. At the beginning of Marc's career, he worked in a lab where moments after cats were killed people would casually ask something like, "Where should we go for lunch?" The cats previously had had parts of their brains removed (or "ablated" in lab-tech lingo) to see how well they could discriminate visual cues. It was because of a cat named Speedo that Marc left this research group. (Researchers weren't supposed to name the cats; they were referred to only as numbers. But Speedo was special.) Marc was appalled by the level of disrespect the researchers showed for Speedo and the other cats on whom brain surgery was performed. Marc also remembers a particular moment when Speedo locked eyes with him, a move that Marc interpreted as Speedo asking, "Why in the world are you doing this?"

Jessica sat for a time on a university Institutional Animal Care and Use Committee (IACUC), which always started its meetings with lunch, so that the review of the first few lethal experiments took place while people ate ham and turkey sandwiches. This cavalier attitude toward animal death is evident, on a broader scale, in the IACUC rating system. Research studies that involve little to no pain or distress but still end in the death of the animal are usually considered Category C, and generate little ethical concern. It is also evident in the concept of "humane end points," which has been developed as a welfare refinement. A humane end point is the "earliest indicator in an animal of pain, distress, suffering, or impending death on the basis of which an animal is killed."[27] In other words, there comes a point in many studies (toxicology, tumor growth, etc.) where the pain and distress of the animal inches past a decency line and killing is the humane response.

What's particularly interesting within the context of animal research is that the "death doesn't harm animals" claim is often couched as a matter of scientific truth. But there is little science to back it up. The claim that death doesn't matter to animals is based on several assumptions: that animals have no awareness of death, no sense of past or future, and no enduring relationships whose loss would cause more than momentary distress. The question of whether animals have an awareness of death remains open, but there are compelling observations of animals mourning the loss of family and friends and performing rituals for them. The most well-known examples of animals who show an interest in the dead are elephants, chimpanzees, and crows.[28] The other two assumptions, that animals have no sense of time and don't form relationships whose loss would cause enduring distress, have been proven false by detailed ethological research. Putting forward the claim that animals cannot be harmed by death because they don't "have a concept of death" is overreaching.

The other main component of the "death doesn't harm" claim is that animals are "dispatched" without fear or pain—so-called humane euthanasia. The confidence with which scientists assert that the deaths of animals in the lab setting are quick and painless is overstated, to say the least. Ongoing research into "evidence-based euthanasia" produces constant surprises and revelations. Every time a new set of euthanasia guidelines is released by the American Veterinary Medical Association, things have changed. Methods once thought to be innocuous turn out, on further study, to cause significant pain or distress and new methods are proposed.

Each and every method of euthanasia has welfare risks—fear, the sensation of being unable to breathe, a feeling of acute stinging pain, the stress of being restrained—that come up before and during the procedure. For example, each of the five most popular methods of killing rodents (CO_2 inhalation, isoflurane inhalation, intraperitoneal barbiturate, decapitation, and cervical dislocation) cause some measure of distress at some point during the "procedure." A consensus meeting on carbon dioxide euthanasia of laboratory animals concluded, for example, that "there is no 'ideal' way of killing animals with CO_2—both pre-fill and rising concentrations can cause welfare problems. . . . If animals are placed into a chamber containing a high concentration of CO_2 . . . they will experience at least 10 to 15 seconds of pain. . . . If animals are placed into a chamber with a rising of concentration of CO_2 they will find it aversive . . . and may experience 'air hunger.'"[29] Air hunger or pain—which would you choose? As veterinarian and historian Larry Carbone says, "Any given species-method combination will have its own profile of welfare concerns."[30] In other words, "humane euthanasia" is currently beyond our technical grasp.

"Noninvasive" Research

The welfare compromises imposed on animals by human researchers run the gamut from the unbelievably barbaric to the relatively benign. On this spectrum of awfulness, research protocols that are labeled "noninvasive" are thought to present no welfare concerns. A "noninvasive" study is basically one that doesn't involve physical injury to the animal. Yet the label is extremely misleading.

Consider imaging studies used in animal research. Imaging studies allow researchers to look at the inside of an animal without cutting her open, and are thus often described as noninvasive. Computed tomography (CT), positron emission tomography (PET), single photon emission computed tomography (SPECT), and functional magnetic resonance imaging (fMRI) are increasingly being used in research protocols that involve tracking the progress of a disease or a treatment, for example, cancers and neurodegenerative diseases. Scientists can gather much of the same information, and sometimes even more, than they might through killing and dissection. It also allows a "view over time." Rather than killing a group of animals at each phase in the development of a disease to study disease progression, researchers instead take imaging studies at these various stages, watching the progression of the disease in the same animals. It allows scientists to increase "the amount of information we get out of the subject," as François Lassailly, a biologist specializing in imaging at the Cancer Research UK London Research Institute (LRI), explains.[31] The animal is treated more like a human patient, becoming what Lassailly refers to as an "ani-patient," undergoing much the same battery of clinical treatments and diagnostic tests as a human patient, and leading to much better collaboration between basic research and its clinical applications.

"Animal handling" aspects of imaging, particularly preparation for imaging and restraint during imaging, can significantly influence ani-patient well-being as well as scientific results. Isabel Hildebrandt and colleagues, for example, reviewed some of the things imaging studies can encompass for ani-patients.[32] They note that a PET scan involves intravenous or intraperitoneal injection of a positron-emitting imaging probe into the animal, for anywhere between five minutes and an hour, and CT scans require high doses of contrast agents. Performing CT, fMRI, PET, and other imaging technologies requires that the ani-patient be kept completely still, and the only way to do this is to use some form of restraint. Typically, full restraint is achieved through anesthesia. Injectable anesthesia agents obviously require injection into an animal, which requires temporary handling and restraint, while inhalation anesthesia requires restraint so that a mask can be placed over the ani-patient's face. A stereotaxic device available for use in PET imaging of the rat brain requires that parts of the device be implanted directly into the animal's skull. With all of the "handling" required by these techniques, it seems a bit of a stretch to call them noninvasive. Furthermore, the animals being imaged will still likely be killed, just at a later end point. They likely also live with progressive disease for longer, increasing the chances that they will suffer.

Even if all that is involved in an imaging study is forcible restraint of an animal's body, and no physical injury or pain is involved, animals can still experience psychological discomfort, and sometimes even extreme distress. Anyone who's ever tried to restrain (by holding in your arms) a pet or a child knows that after a few minutes they will squirm, struggle, cry out, and get increasingly distressed if not set free. Just as standards for how long farmed animals can be expected to reasonably stand in a truck or train traveling to slaughter without rest, food, or water (28 hours) seem cruel, so too do the expectations for restraint of ani-patients. Here is a standard guide, as found on the Research Animal Resources website for the University of Minnesota: "Animals should be released from restraint devices at least daily and allowed unrestrained activity to prevent muscle atrophy and skin necrosis."[33] Once a day. For research studies that extend over the course of a number of weeks, restraint of an awake, fully conscious animal is beginning to sound worse than a quick death.

All in all, like the word "humane," we shouldn't take the term "noninvasive" to mean "without effects on animal well-being" or "without moral compromise."

Degrees of Freedom and Consent

One of the central considerations in this book is how to improve animal well-being by increasing the degree of freedoms that individual animals experience. Data from various research studies, including notably Martin Seligman's research on learned helplessness, show that inability to control one's environment exposes animals to stress and negative emotions.[34] As Seligman and his colleagues reported,

When a normal, naïve dog receives escape/avoidance training in a shuttlebox, the following behavior typically occurs: At the onset of electric shock the dog runs frantically about, defecating, urinating, and howling until it scrambles over the barrier and so escapes from the shock. . . . However, in dramatic contrast . . . a dog who had received inescapable shock while strapped in a Pavlovian harness soon stops running and remains silent until shock terminates . . . it seems to "give up" and passively "accept the shock."[35]

And, as a corollary to learned helplessness and despair, research data strongly support the idea that the well-being of animals is enhanced when they have the ability to control certain aspects of their environment. To put this in the simplest possible terms, freedom feels good and the lack of freedom feels bad.

Animal research obviously fails the freedom test by flunking the first requirement of human-subject research: Do you have consent? If we asked animals first if we could cut them open, give them cancers, sever their spinal columns, or inject them with HIV, they would scurry away as fast as their legs, wings, fins, or tails would carry them. This should, of course, give us pause. But, given that invasive research on animals is going to continue at least in the short term, are there things we could be doing to increase control and choice, to gain at least some small measure of consent from our animal subjects? Yes, there are.

We wrote in chapter 3 about increasing a sense of control through voluntariness—allowing captive animals to take control of certain aspects of their environment by choice, not coercion. Instead of forcing cows to milk at an allotted time, why not let them decide when their udders are in need of relief? An analogous move in laboratory-animal welfare is the use of positive-reinforcement techniques such as a food reward to train animals to voluntarily participate in some aspects of research. Consider, for example, venipuncture in primate species. Blood draws are one of the most common procedures carried out on primates, and they are known to be stressful. The animals typically need to be restrained, either pharmacologically, through sedation, or physically, through the use of special restraint cages. Both of these methods of restraint can alter levels of cortisol in the blood, throwing off the accuracy of the sampling results. Several species of primate have been trained to voluntarily present an arm for a blood draw. For example, a study by Kristine Coleman and colleagues described training a group of Rhesus macaques to place an arm in a "blood sleeve" and stay quiet while blood is drawn. Three-fourths of the monkeys learned to volunteer.[36] Similar training with chimpanzees has been even more successful. Researcher Hal Markowitz writes, "Today it is not unusual to see a competent veterinary technician and keeper working together with an ape that has been trained to tolerate the discomfort associated with inserting a hypodermic needle."[37] Markowitz was speaking about apes in zoos, but the principle remains the same in the laboratory setting. Of course, putting one's arm out for a needle is only one small voluntary act within a much larger coercive one, but better than nothing.

Eliza Bliss-Moreau and her colleagues offer another variation on voluntariness, again with primates. Sometimes an experimental design will require animals to be awake and aware during testing, "necessitating some form of restraint."[38] This is often true, for example, in research studies of physiological and cognitive processing. One of the most common restraint devices for primates is called a "primate chair." The plastic variety looks like a large clear box with a plastic chair inside. The animal sits in the chair, his body enclosed within the plastic cage and, in some models, held still by a grate or "squeeze panel" that can be moved forward. The animal's head sticks out a hole in the top, like a jack-in-the-box, and the animal is immobilized. The standard way to get a monkey into a chair is the so-called pole-and-collar method. Monkeys wear metal collars to which long poles can be attached. Handlers keep their distance from the monkeys and use physical force to move the animals, who naturally struggle to get away. This method of transfer is stressful for the animal, and stress, of course, can impact the quality of the data.

As an alternative to the pole-and-collar method, Bliss-Moreau's team trained a group of monkeys to participate in "voluntary yoking." The "cooperative training" involved a pastiche of positive reinforcement, desensitization, and negative reinforcement. It took the team an average of fourteen days to teach the monkeys to submit to voluntary yoking. Although the claim made in their first sentence—"It is sometimes necessary for nonhuman primates to be restrained during biomedical and psychosocial research"—is standard welfarist fare, they nevertheless offer a refinement to the use-of-force approach. But their study also highlights, again, how very narrow the slice of voluntary action is. The monkeys are going to be placed in the chairs, regardless. If they cooperate, the experience may be slightly less distressing for them.

Yet another example is cited by Viktor and Annie Reinhardt in their book on enrichment for rodents in the laboratory setting. Gastric intubation for oral drug administration is a procedure routinely carried out on rodents and rabbits and is known to be extremely distressing. "The animal is usually exposed to two or three humans who apply forceful restraint/immobilization during an extremely uncomfortable, life-threatening, often injurious and sometimes even deadly procedure."[39] As an alternative, several researchers trained a group of socially housed Wistar rats to cooperate during oral administration of two anti-inflammatory drugs. The rats were encouraged to develop a taste for chocolate, which was very easy because rats love chocolate. Handlers gently held each rat and placed a pellet of chocolate in his mouth using a fine-gauge needle, working on the same principle dog owners follow when hiding a pill inside a treat rather than shoving it straight into the dog's mouth. The rats soon learned to eagerly participate in the ritual, and after eight days of training were happy to take a chocolate pellet laced with the test drug.[40]

In each of these examples, animals likely have an increased sense of control over their environment, and the stress associated with a painful and scary procedure has been reduced. Of course, participation in research is hardly voluntary. We ask nicely the first time, but opting out is not an option and force is the ready standby. And unfortunately, using positive-reinforcement techniques with animals is time- and labor-intensive, and only a small number of animals are lucky enough to live in labs where technicians or researchers will make the commitment to train their subjects.

For comparison, consider some animal research that verges on being truly voluntary and consensual. Several scientists studying dog cognition have been "inviting" participation by companion dogs. Brian Hare, Alexandra Horowitz, Adam Miklosi, and several others have been developing experiments in which dogs get to play games or solve puzzles.[41] The research is totally noninvasive, fully voluntary, and fun for the animals. Neuroscientist Gregory Berns has studied dog cognition using individuals who were trained to lie still for fMRI scans. There were no catchpoles, no restraint devices, just positive-reinforcement training. Furthermore, if dogs didn't want to participate, they didn't have to and could opt out. The research subjects were companion animals who were part of human families, and not caged in a research laboratory. After the fMRI, or after their refusal, the dogs were not put back in cages, nor were they euthanized. Instead, they returned home.

In some situations, such as veterinary care of companion animals, painful or scary interventions, such as an injection of antibiotics or surgery to remove a tumor, are sometimes "forced" on animals for their own good. These situations are inherently different, because the procedures are truly for the animal's benefit, so some coercion or lack of consent may be justified. That said, there is a value in making these experiences as "fear free" as possible.[42]

Again, it is important to emphasize that increasing voluntariness is not only good for animals but for science, since stress alters physiology and can confound data. Bruno Preilowski and his colleagues were doing research on brain mechanisms in rhesus macaques. He designed a protocol where the monkeys had control over the experiment and could choose when to participate. His results, he claimed, were more significant than in earlier protocols in which the trials were forced on the monkeys.[43]

Autonomous Animals

When bioethicists talk about the ethics of research on human subjects, the key value they urge policymakers and researchers to uphold is respect for autonomy. Autonomy, from the Greek *auto* (self) + *nomos* (law) is the freedom to be self-governing, to think and act independently and without coercion. Within the context of medicine, in particular, it is the capacity of a rational individual to make a free and informed decision about his or her treatment and care. It is rare to hear the language of autonomy used in relation to animals, perhaps because of a long-held prejudice that animals are not rational. Yet as we learn more about the cognitive capacities of different species, it is inevitable that the question will come up: Should we seek to respect the autonomy of animals, as we do people?

In a pioneering essay, bioethicists Tom Beauchamp and Victoria Wobber argue that chimpanzees can act autonomously, although the psychological mechanisms underlying the behavior are not the same as those of humans. Beauchamp and Wobber review the current literature on chimpanzee cognition and conclude that chimpanzees satisfy the two basic conditions of autonomy: liberty (the absence of controlling influences) and agency (self-initiated intentional action).[44] Chimpanzees, they argue, act with understanding, intention, and self-control, and "make knowledge-based choices reflecting a richly information-based and socially sophisticated understanding of the world." Within larger groups, wild chimpanzees are free to choose different grooming and play partners, based not on some "genetic imperative" but on their assessments of whom they like or dislike and whose back might be worth scratching.

"Autonomy" is a psychological mechanism; "respect for autonomy" is a moral principle. Beauchamp and Wobber focus attention on the psychological mechanism, leaving open what this means in moral terms. Still, in raising the possibility of species-specific autonomy, they set the stage for a significant recasting of our moral obligations to animals, particularly within the realm of research.

Captivity as Imprisonment

Even if we provide animals with an enriched environment and even if we increase the voluntariness of our research practices, a fundamental violation still remains intact: the violation of the animals' liberty. Liberty is understood as the state of being free from oppressive restrictions on one's way of life and behavior. Among these oppressive restrictions are physical constraints, confinement, servitude, and forced labor. It may seem radical to use the term "liberty" with regard to animals; indeed, media reports often frame the agenda of the animal-liberation movement as radical. But when we simply think about what animals lose when we lock them up and toss away the key to freedom, applying the term "liberty" doesn't seem so radical after all. Nearly everyone can agree that animals value their liberty. Why else would they struggle against restraint? As the University of Minnesota's Research Animals Resources website explains to animal technicians and researchers: "Handle animals firmly. The animal will struggle more if it sees a chance to escape."[45]

World-renowned primatologist Sue Savage-Rumbaugh, among others, has tried to articulate just what it is that animals lose when they lose their liberty. In relation to the great apes in particular, Savage-Rumbaugh challenges the standard welfarist approach, which equates providing an enriched environment with animal happiness, and which fails to look beyond the superficial needs of the animals. "Humans assume that an adequate captive environment for apes consists of simply a few acres with trees; a varied diet; enrichment items; and individuals documented to play, groom, and produce offspring with one another."[46] This "happy ape world" is an illusion, she says, and mirrors what we want apes to be, not who they actually are. We "characterize them as clever, self-centered, emotional beings who are on the brink of humanness but still sufficiently different from humans so that our concerns for their well-being need not overlap with the concerns we traditionally raise regarding human welfare (the rights to personhood, independence, dignity, free speech, freedom from oppression, the development of one's potential, the choice of a mate, the rearing of one's offspring)."[47]

Savage-Rumbaugh notes the employment of two separate vocabularies to discuss human and nonhuman mental states and this, she argues, leads us to different understandings of what constitutes well-being. For example, when talking about apes, we use the biological terms "male" and "female." We avoid using terms ("husbands," "wives," "children," "partners," "friends," "teenagers," "parents," "uncles") that imply social relationships and normative group behaviors.[48] Is the use of separate vocabularies justified? On scientific grounds, she argues, no. We strongly agree. The continuities in behavior and neurophysiology are vast, compared to the differences. The use of a separate vocabulary suggests that "apes have no culturally constructed social roles." Researchers thus think nothing of exchanging an ape here for an ape there, manipulating the membership of groups at different facilities as if exchanging tokens, not children, mothers, uncles, allies. All the while, they fail to recognize the psychological effects of this kind of social disruption. Savage-Rumbaugh writes, "The inner minds and lives of apes exist on a plane of invisibility for humans, and we do not see the pain we cause them with our lack of understanding."[49] As she says,

> We wish to create good feelings in ourselves by giving objects, trees, and space to our captive apes, but we continue to take from them all things that promote a sense of self-worth . . . No captive environment requires cooperation or group co-

ordination, so captive apes have no need to construct and maintain a cultural stance toward mutual group action across significant spans of time. Their captive environment negates the possibilities of travel, kinship structure, roles within the group, group-based mental worlds, and constructs of cultural realities.[50]

"Why," she finally asks, "do we imprison apes?"[51] She is one of the few scientists to use this language. If more did, it would become obvious to them that the animals they are studying really are prisoners, whether they are in cages or larger enclosures. As Jane Goodall has said, "The least I can do is speak out for the hundreds of chimpanzees who, right now, sit hunched, miserable and without hope, staring out with dead eyes from their metal prisons. They cannot speak for themselves."[52]

The Fourth and Fifth Rs: Refusal and Rehabilitation

In one small corner of the animal-research world, something revolutionary has been happening. A movement is afoot to completely free certain animals from captivity and, more particularly, from use in biomedical research. This revolution represents what are sometimes referred to as the fourth and fifth Rs: refusal and rehabilitation (expanding on the three Rs, reduction, replacement, and refinement, discussed previously).[53]

Refusal and rehabilitation efforts are focused, for now, on the great apes, who are gradually being phased out of research and re-homed in sanctuaries. As of this writing, a number of countries, including Germany, Sweden, the Netherlands, the United Kingdom, and New Zealand, have banned or placed severe restrictions on the use of great apes in biomedical research. Austria, which banned the use of great apes more than a decade ago, goes a step further and also forbids the use of lesser apes. The United States is a notable holdout, with about twelve hundred great apes still being held in research facilities. Yet even the United States is moving in the direction of a ban. In 2011 the Institute of Medicine concluded that most biomedical research on chimpanzees is unnecessary, and in 2013 IOM recommended a phasing out of great ape research and the eventual retirement of all chimpanzees currently being held in labs.[54] In June 2015 research chimpanzees were declared to be endangered under the Endangered Species Act, ending nearly four decades in which captive chimpanzees were denied the same designation as wild chimpanzees, who have been listed as endangered since 1976.

This new designation will make their continued use in biomedical research even more unlikely.[55] Apes who are currently housed in labs will be rehabilitated, which is a nice way of saying that they will live out their lives in a white-collar prison. They will have space, enrichments, even human affection. An effort will be made to give back some of what they have lost, which is sometimes physical and sometimes mental and often a combination of the two. But they will never have what we took from them in the first place: a life in the wild. Nor can they ever be returned to the wild, where they would be unable to survive.

One of the driving forces behind the phaseout of primate research is the Great Ape Project (GAP). Founded in 1993, GAP is an international group of primatologists, anthropologists, biologists, philosophers, and others who advocate a United Nations Declaration of the Rights of Great Apes that would confer basic legal rights on the great apes: chimpanzees, bonobos, gorillas, and orangutans. The rights suggested are the right to life, the protection of individual liberty, and the prohibition of torture. The organization also monitors individual great ape activity in the United States through a census program. Once rights are established, GAP would demand the release of great apes from captivity. Attorney Steven Wise and his colleagues at the Nonhuman Rights Project are also continuing to work very hard to get courts to grant rights to great apes.

The phasing out of great apes in research is part of a broader trend, and eventually refusal and rehabilitation may be expanded to other species. A consensus has been gradually forming around the idea that some species of animals simply cannot ethically be subjected to imprisonment or experimentation, either because of their cognitive and emotional sophistication or because of their special place in our moral framework or both. For example, it may be that dogs and cats will increasingly be seen as inappropriate lab objects, not only because of their intelligence and emotionality but also because of their special bond with humans, at least in those parts of the world where the keeping of dogs and cats as companion animals is common. The kinds of questions Savage-Rumbaugh is asking about apes can and should be asked about other species, and we can endeavor, with the help of science, to elevate the inner minds and lives of other animals above the "plane of invisibility" and become cognizant of the pain we cause through our lack of understanding. The science of animal well-being needs to include this kind of careful, species-specific exploration of animal capacities, and it is likely that taking the science seriously will make refusal and rehabilitation an increasingly compelling and realistic goal.

Better Welfare = Better Science

Russell and Burch warned that the wages of inhumanity are "paid in ambiguous or otherwise unsatisfactory experimental results." One of the things that animal researchers have discovered, over and over again, and which is a boon to the research animals themselves, is that better welfare equals better science. This has been the main impetus for improving lab-animal welfare. Nevertheless, good science and good welfare are still a long way off.

We've already noted that stress affects physiology and can thus affect the quality of data. But there is still a great deal to learn about exactly what precipitates stress in captive animals, because sometimes things as seemingly innocuous as the color of the cage can influence how animals feel. Furthermore, stress indexes and emotional states are not well correlated, so we can't assume that an animal with normal physiological parameters is unstressed.

Animal sciences professor Joseph Garner writes, "It is useful to think of behavior as an organ, which is integrated with the biology of the whole animal . . . behavior is intimately involved in homeostasis."[56] In other words, alterations in behavior have effects on physiology, which in turn have effects on the validity, reliability, and replicability of scientific outcomes. Scientists ignore behavioral pathologies such as stereotypies at their peril. Some researchers seem to think that utterly barren environments are best, because they are all the same and you don't introduce any variability into your study. All the animals will be doing the exact same thing: nothing. Barren cages are also cheap and easy, which is an added bonus for the harried researcher. Yet Garner argues that the opposite is true: enrichments may improve the validity, reliability, and replicability of results by reducing the number of abnormal animals introduced into a given study. Nevertheless, the need to introduce variables in the form of enrichments simply adds another layer to the challenge of getting "clean" data.

There are myriad ways in which compromised welfare results in compromised science. Just recently, for example, a group of researchers at the US National Institute on Aging in Bethesda, Maryland, expressed concern that many rats and mice used in experimental studies are so overfed they may die prematurely, and that such premature deaths may skew data collection in areas as diverse as immune function, cancers, and neurological disorders.[57] A *New Scientist* article, "Too Stressed to Work," cites research on rats housed in stressful conditions. Rats "show an inflammatory response in their intestines accompanied by leaky blood vessels. . . . As a result, the gut's defence barrier breaks down, leading to chronic inflammatory conditions such as 'leaky gut.' This inflammation adds uncontrolled variables to experiments on these animals, confounding the data."[58]

Another recent study, published in *Trends in Cancer*, noted that even something as subtle as air temperature in the laboratory can induce stress in animals and can, in turn, affect data. Immunologists Bonnie Hylander and Elizabeth Repasky have been investigating the effects of cold stress on the mouse immune system. Labs are often kept fairly cool, since researchers wear robes, gloves, and masks and can become quite warm while they are working. Yet Hylander and Repasky found that the cold temperatures also affect the mice, whose heart rate and metabolism change as their bodies try to generate heat. Tumors grow and metastasize more quickly, and respond less well to chemotherapy in mice who are cold than in mice whose bodies are warm. The researchers are concerned because reported data don't generally take into account ambient temperatures in the laboratories where the research was conducted and the data may, therefore, be misleading.[59]

Indeed, and this is the frightening part: there are likely lots of ways in which data are skewed of which we aren't even aware. All the while, this is not only bad for the animals but bad for people. The interactions of poor welfare, unseen sources of stress, and the nuances of the parameters being measured, not to mention subtle differences in behavior and physiology that each individual animal brings to the table, all combine to create a perfect storm and we must be very careful to emerge from this storm with reliable scientific data.

And even when all welfare standards have been met and rules are followed to a tee, animal welfare still falls short. Standard operating procedures such as restraint and confinement impose significant harms that cannot be offset by enrichments. And we still fail to provide animals with what they really need and want, which is freedom. Even the best welfare is not good enough.

Moving Beyond Welfarism

There is no denying that some animal research has greatly benefited humans, and continues to do so. But this does *not* commit us to an indefinite future of animal research. As the Institute of Medicine notes, "past use fails to predict future necessity."[60] In fact, getting stuck in the "we wouldn't have X, Y, or Z treatment if it weren't for animal research" is counterproductive, because it keeps us focused on the past when we need to be looking toward the future. Humans are innovative enough that we surely can develop a broad range of nonanimal models that surpass, in efficacy, our current state of the art. Great progress has been made, and more money and research energy should be focused in this direction. If animal research were to slow to a trickle, there is no doubt that innovation in alternatives would flourish.

One example of a promising alternative is the Vanderbilt-Pittsburgh Resource for Organotypic Models for Predictive Toxicology, which is developing toxicity tests based on three-dimensional human cell cultures. This is part of the Environmental Protection Agency's Tox21 initiative, which aims to reduce reliance on animal tests and, by 2018, to eliminate the use of animals in toxicology. These 3-D human cells are likely to produce much more reliable data, because animal and human cells often respond differently to the same chemicals, and animal tests have provided some misleading information about toxicity.

The final chapter in Russell and Burch's *Principles of Humane Experimental Technique* speaks to their concern over long delays in the application of existing knowledge to the improvement of experimentation. In "The Factors Governing Progress" they cite, as examples of the knowledge-translation gap in the late 1950s, the use of incorrect or inapplicable statistical models. "Delays of this kind may be regarded as a sort of inertia, or rigidity, the maintenance of a habit . . . long after information is available for its correction."[61] Often such delays are simply a result of isolation or lack of communication. This is the real shame of animal research: when the suffering of animals leads to nothing more than piles of bad science. And unfortunately, a good deal of animal research is just bad science.

The quality of scientific output has been under increasing scrutiny lately, with accusations of researchers massaging data or publishing results that cannot be replicated. There is tremendous competition for funding and increasing pressure to produce "data," so much so that the design and implementation of research studies is often shoddy and aimed at getting quick results. Scientists study questions for which ample data already exist, such as the proposal by a group at the University of Wisconsin to repeat Harry Harlow's horrific experiments on the effects of maternal deprivation on the psychosocial development of young Rhesus monkeys. The proposal incited so much public outrage that the research group agreed not to repeat the maternal-deprivation aspects of the study.[62] Too many studies, often of the blindingly obvious, are poorly designed, generate poor data, and cost the lives of millions of animals. This is offensive beyond measure to those animals who suffered and lost their lives, just as it is an insult to cows and pigs and chickens when their flesh is simply tossed into the trash.

What kinds of things can we do to improve the well-being and freedoms of individual animals used in research and testing? We can *reduce* the number of animals being used by making sure that all science involving animals is rigorous, impeccably designed, carefully vetted, and absolutely necessary; we can *replace* animals with computer models and other innovative nonanimal alternatives; and we can *refine* experimental protocols so that individuals suffer fewer harms and have greater control over their own lives within the context of their confinement. And we can also work toward going beyond the Three Rs to the fourth and fifth, *refusal* and *rehabilitation*, for an expanding number of species. The science of animal well-being moves us beyond welfarism by taking seriously what we know about animals and what they need, by making the enhancement of real freedoms for individual animals, especially freedom from exploitation, our first priority.

The ethical back-and-forth can go on interminably as long as we stay in a welfarist mind-set and work to repair instances of "unnecessary suffering" while failing to challenge the underlying violence. We can challenge the Claude Bernards of the world, who will try to justify whatever research they feel "needs" to be done. We first have to recognize that the denial of freedom to these millions of individual animals is wrong. We must turn our full energies to a transition toward a new paradigm that causes the least harm to all involved, which may entail not allowing for certain sorts of research or stopping some ongoing projects in their tracks, such as those trying to prove the obvious or inconsequential, those that are generating poor or inessential data, and those that involve social deprivation, food or water deprivation, and learned helplessness. Researchers—from those already well established to those just starting out in their careers—need to keep in mind that better care leads to better science, and that this is a win-win for everyone involved.

CHAPTER 5

Zooed Animals

RARE WARTY PIGS ARE LOST WHEN MALE
EATS HIS ENTIRE FAMILY AT BRISTOL ZOO
A zoo lost some of its most endangered animals when a male warty
pig ate his entire family and a rare monkey was eaten for lunch by
hungry otters. . . .

The most gruesome incident came after Manilla, a female
Visayan warty pig was joined by her partner Elvis last year.

Staff said they hoped he would "take a shine" to her and they
would "become proud parents."

But when she unexpectedly gave birth to an extremely rare
piglet, Elvis ate it before turning on his mate, who had to be put
down due to her injuries.

Two weeks later, an endangered golden lion tamarind [sic]
monkey escaped and fell into a pond where it became trapped
and was eaten by American otters.

Just a week ago three rainbow lorikeets, usually found in
Australia, escaped through a hole in their cage [and] flew off.
One is still on the loose.

—*Telegraph* (London), February 5, 2015

The plight of animals in entertainment has gained unprecedented
public attention over the past several years, and much of the con-
sciousness-raising has occurred by way of a particular orca whale
named Tilikum, known by his nickname, Tilly. Tilly was captured
near Iceland in November 1983. When he was only two years old, he
was torn away from his family and his ocean home. After a number
of years of being transferred from one aquarium to another, Tilly
was finally acquired by SeaWorld San Diego, and became one of the
star attractions and moneymakers for the theme park. But the years
of captivity and maltreatment took a toll on Tilly, and he started

behaving erratically. He eventually killed one of his trainers, in front of a horrified audience. The details of Tilly's tragic life and fateful end were beautifully captured in a documentary called *Blackfish* (2013). By weaving together ethological details about the cognitive, emotional, and social lives of orcas in the wild with a catalog of the abuses and deprivations experienced by Tilly, the film leaves the viewer in no doubt that SeaWorld is a living hell for these sensitive and intelligent creatures, who go crazy and must be pumped up with psychoactive drugs like Valium to control their behavior.[1]

SeaWorld, for its part, has seen ticket sales plummet. In March 2016, SeaWorld announced that it will end its orca breeding program, won't obtain new orcas from other sources, will begin replacing its theatrical orca shows with shows that exhibit the whales' natural behaviors, and will have no orcas at all in any new parks around the world. Further, SeaWorld has pledged to invest millions of dollars for the rescue and rehabilitation of marine animals. (And in a nod to animal welfare more broadly, it also pledged to use cage-free eggs, gestation-crate-free pork, and sustainably sourced fishes at SeaWorld venues, and offer more vegetarian and vegan options.)[2] This is a very good beginning, and we can look forward to the day when venues like SeaWorld, including terrestrial zoos, morph into sanctuaries in which the animals' lives are put first and foremost. It is vitally important to keep working toward these goals. As the longtime animal advocate Gretchen Wyler once noted, "Cruelty can't stand the spotlight."[3]

So how do animal well-being and freedom fare within the various entertainment venues that center on animals? Before we get to our main focus in this chapter—zoos—consider a few instances of animals being used for entertainment.

Last Chain on Billie: Bridging the Empathy Gap

Billie, as she became known, was a majestic Asian elephant who was captured and shipped to America as a baby. She was taught to perform circus tricks such as standing on a small tub and balancing on one foot. Audiences clapped and marveled at her skills, but they didn't know about Billie's miserable behind-the-scenes life as she was hauled across the country and kept chained for hours on end when she wasn't performing. Billie suffered enormous physical and psychological trauma but she finally got a lucky break when she was rescued and sent to a sanctuary for performing elephants in Tennessee. While she would never again be chained or beaten, her abuse had taken its toll, and Billie refused to interact with the other elephants or to allow anyone to remove the chain around her leg. For five years, Billie wouldn't forget her past, and the chain remained. Finally, Billie allowed caregivers to remove it with a bolt cutter, from the other side of a fence. As Carol Bradley writes in her wonderful book *Last Chain on Billie*, "It was almost as if Billie realized what they were attempting to do. The expression on her face softened and she stopped swinging at the fence. She lifted her foot again, this time higher than before, pushed up against the bars of the fence and rotated her ankle first one way and then another." Billie didn't give any indication that she understood the significance of being freed from the chain, and "headed out to the sand pile to wake [her elephant friend] Frieda from her nap."[4] When the chain finally fell from her leg, she "picked it up with her trunk, then dropped it and walked away." Billie, Bradley writes, "had better things to do."

Circuses involve humiliation, punitive training techniques, and poor living conditions for captive animals. Rodeos and dog- and cock-fighting make spectacles of animals through overt violence. Dog- and horse racing are exploitative and often involve physical injury to the animals. Even venues in which humans can interact with "wild" animals and where the animals are not subjected to overt abuse can be harmful. Opportunities to swim with dolphins or pet wild tiger cubs involve a loss of freedom for the animals and a disruption of their lives, for the purpose of momentary human delight.

Billie's story is hardly unique, and it captures just what happens to numerous entertainment animals whose hearts and spirits are broken as they are abused "in the name of entertainment." A very useful thought experiment for bridging the empathy gap is to ask, "Would I do it to my dog?" This question brings home not only what entertainment animals endure, but also how our companion animals, who aren't any more sentient than the animals used in entertainment, nevertheless enjoy a higher status. Most people would never imagine letting a dog be treated in such disrespectful and abusive ways.

The Case of Zooed Animals

Let's now turn our attention to animals displayed at zoos and aquariums, where humans are simply observers, where an attempt is made to provide animals with suitable, even naturalistic, habitat, and where veterinary care and nutritious food are provided to all residents. Are animals living under these conditions happy and content, as zoo brochures would have us believe? The simple answer is no, they are not. Some zoos, particularly the thousands of roadside attractions, are shockingly mismanaged, and animals suffer from neglect, poor care, small, barren cages, and no attention to their species-specific or individual needs. But there are many zoos that seek to maintain the highest standards of care for animals, often through the application of animal welfare science. These are the focus of our attention. It is easy to see the problems for animals at the worst places; it is more interesting to take a look at the better institutions and what they are doing to improve the lives and well-being of animals. We can see how welfare science can make an enormous difference in the lives of animals. We can also see why good welfare is not and can never be good enough and how the animals on display suffer from huge losses of freedom.

According to the Association of Zoos and Aquariums there are more than 10,000 zoos around the world. In the United States, as of September 2015, there were 230 AZA-accredited zoos and an estimated 2,400 "animal exhibitors" licensed by the USDA. (This number does not include individuals who own one or more exotic animals as pets.) AZA accreditation means that an institution has met certain standards for animal care and management, paying attention, for instance, to proper housing, nutrition, and social groupings.[5] More than ten thousand different species are held in zoos around the world.

In zoos, welfarism and welfare science are hard at work. As with animal welfare on factory farms and in laboratories, science casts its sheen over our interactions with animals in zoos. A special issue of the *Journal of Applied Animal Welfare Science* (*JAAWS*) devoted to "issues in zoo animal management" suggests that welfare problems in zoos are issues of management, not ethics, and that good, scientifically informed management strategies can resolve welfare concerns.[6] As in other realms, the welfarist paradigm controls the dialogue about animals. It focuses public and scientific attention on what we can do better for animals within the current paradigm. But it never pushes the questioning beyond the doors of the animals' cages. Zoo workers justify the keeping of animals in captivity by appealing to the welfarist paradigm, saying that animals in captivity are safe and comfortable and better off, in some respects, than they would be in the wild. Yet as ethicist Koen Margodt puts it, zoos are essentially "welfare arks" in which animals are collected, purportedly to save them from extinction, but where human interests are put before the interests of individual animals.[7]

Indeed, although zoos are open to the public, and sometimes even funded by local initiatives and taxes, what goes on behind the scenes is hidden from the public, unless and until problems become so severe that the bad news spills out, which it often does. The Smithsonian's National Zoo in Washington, DC, for instance, was recently the subject of an investigation into mismanagement and animal welfare problems related to the zoo's Cheetah Conservation Station. The zoo decided to double the number of animals in its CCS exhibits, but without increasing the amount of space allocated to the animals or preparing in advance where and how the animals would be housed. The results, not surprisingly, were tragic. Because there was no exhibit space ready for them, two newly acquired hornbill birds were kept in a shack for seven months. When zoo workers complained to management, the birds were put into the wallaby exhibit, which stressed out the wallaby, who bloodied his nose on the enclosure wall when he tried to run away. Holly, one of two newly acquired red river hogs, became so malnourished she died. A dama gazelle and a kudu both died after running into barriers in their enclosures and breaking their necks. A young Przewalski's horse also died after he broke his neck in a cage at the zoo's Conservation Biology Institute in Front Royal, Virginia. Przewalski's horses are listed as endangered animals, and in the recent past were listed as critically endangered, so this loss is not only tragic for the individual, but also for this species as a whole.

Zoo officials insisted there were no problems, and the animals were doing just fine. Steve Feldman, a spokesman for the AZA, said in a news interview that the zoo had recently been evaluated and had met the AZA's rigorous criteria for animal welfare, and that the events at the National Zoo were just natural. "The circle of life that occurs for all living things repeats itself in zoos on a daily basis," he said. "Just as with humans, we notice more when it happens to celebrities, we tend to take more notice when these things happen at a place like the National Zoo because their zoo animals are the celebrities of the animal world."[8]

The introduction to the *JAAWS* special issue on management of zoos, by Cynthia Bennett, tells us that zoos and aquariums are committed to enhancing the welfare of nonhuman animals in captivity. Bennett admits, as does nearly everyone writing in the zoo-animal welfare literature, that our techniques and practices are not ideal and that animals are suffering. Animal welfare science was developed within the context of agricultural animals, and it isn't all that easy to transfer welfare considerations from one venue to another, because the goals and focus are different.[9] Within agriculture, the focus is on the productivity of animals over a very short time period. This doesn't translate seamlessly into a zoo setting, where animals are expected to live out their natural life span and where the goal is to keep them healthy, not necessarily to make them fat. Furthermore, welfare science in agriculture is focused on identifying and responding to poor welfare states, and not to actually optimizing happiness or quality of life. And finally, zoos generally forgo common domestic animals like pigs and cows, about whose welfare we have a great deal of information, and instead hope to house exotic, endangered, or charismatic species about whose welfare in captivity we know very little to nothing. Zoo administrators and managers simply can't know what each of the thousands of different species of animal need and want within the captive environment. As a result, not all of these various creatures "are managed in fully evidenced-based, optimized ways."[10]

Bennett concludes, "Acceptable standards and best practices remain elusive and are often subject to debate."[11] Yet she takes for granted that holding animals in captivity is perfectly acceptable and normal. Welfare science deflects attention away from moral concerns over animal captivity, because it stays focused on immediate welfare challenges and provides a nice veneer of scientific and thus ethical acceptability to the overall endeavor. When Bennett poses questions that might shape future discussion, she asks whether we might rethink the accepted rules of facility design "that give people more room and freedom to move than the captive animals they come to see." This is a great question, because it exposes the irony and the fundamental insult of zoos: animals have had their basic freedoms stolen and are viewed as commodities, to feed human pockets.

WHAT KINDS OF VIOLATIONS OF FREEDOM DO ZOO ANIMALS EXPERIENCE?

The basic welfare problem of zoos, and the violation that causes the most misery, is the loss of the big F, Freedom. Attempts to provide bits and pieces of the Five Freedoms are of no consolation to an animal who has lost his most cherished possession.

WHAT DO ZOOED ANIMALS WANT?

This is easy, and we don't need any preference studies to figure it out. They want to be free from captivity. Whether they are born into captivity or "born free," wild animals "want" to live in a setting in which they can engage the repertoire of evolved behaviors that define them as a species.

ZOO MANAGEMENT: WELFARE FOR PROFIT

In some respects, zooed animals are like farmed animals, because every aspect of their lives is managed by their human keepers. In welfarist lingo, this is "zoo animal management." Indeed, running a successful zoo is much like running a successful business, balancing the costs of procuring and managing inventory against the profits from sales.

Zoo animals are carefully bred using studbooks and genetic analysis; social groupings are manipulated; animals are fed and watered at set times; deaths are scheduled and orchestrated by veterinarians or zoo managers; reproductive cycles are watched, sometimes controlled; birth is carefully managed; at some zoos, newborns (the core genetic inventory) are taken from their mothers and killed if not needed, or sold off, if necessary to optimize the zoo's holdings and maximize profit. The costs of providing for animal welfare are always balanced against profit and optimization.

As in the other animal-centered venues we've discussed so far, money is one of the key drivers of zoo-animal welfare. Zoos have discovered that good welfare is more profitable, at least on the whole, than poor welfare. The one-hundred-million-odd people who visit zoos each year in the United States, for example, want to view animals who are active and engaged. It is much more fun to watch a lion stalking past the viewing window than to watch a lion sleeping or pacing back and forth, and far more entertaining to watch penguins dive and swim than to see them huddled on a fake glacier in a refrigerated cage.

Just as welfare of food animals is measured in productivity (number of eggs, liters of milk, quantity of muscle), so too is the welfare of zoo animals often measured not by how happy they are, but by how well they are serving the needs of zoo administrators and the public and how well they are generating money. What are the measures of welfare success used by zoos? One is fecundity; the more fertile a given animal the better they are thought to be doing. There is some science behind this, of course. Stress has a pronounced effect on reproduction. In particular, the stress hormone cortisol suppresses the secretion of reproductive hormones and can disrupt estrus cycles. So, poor reproduction is a sign of poor welfare. The corollary is assumed also to be true: good reproduction equals good welfare. And reproduction is one of the key drivers of a healthy zoo industry. A successful breeding program allows a zoo to build the "captive gene pool" of a given species, which the zoo industry can then use to further replicate animals. Zoos make money selling animals to other zoos; and live breeding animals—itinerant sperm donors—or their semen are traded around among zoos, as zoos seek to create the best collection. It is a great big game of musical animals and musical semen, though not a game that's very much fun for the animals involved. (Marc first heard the term "musical semen" used by Julie Woodyer of Zoocheck Canada.)

One of the more nettlesome management dilemmas faced by zoos is the best way to deal with overly successful breeding and limited zoo space. And on this issue, American and European zoos take different approaches. In American zoos, the preferred approach is to make active use of contraceptives. All manner of animals, from chimpanzees to small rodents, are fed hormones to control ovulation and prevent pregnancies. The upside is that unwanted offspring don't have to be killed. The downside is that many animals never get to engage in some of the most basic natural behaviors: giving birth and raising young. European zoos are less keen on denying their zoo animals these fundamental experiences, and will often allow animals to mate and bear offspring. As Bengt Holst, director of conservation for the Copenhagen Zoo, said in an interview, "We'd rather they have as natural behavior as possible. We have already taken away their predatory and anti-predatory behaviors. If we take away their parenting behavior, they have not much left."[12] Once the young reach the age at which, in the wild, they would become independent of their parents, they are removed and, if they aren't needed for future breeding, they are killed.

Life span is another measure of good management/good welfare. Zoo managers often say things like, "Look! Our animals live even longer than they would have in the wild, which means they are safe and happy." But for the animals themselves, of course, quantity isn't necessarily the same as quality. There is the huge issue of having to live one's entire life in captivity. But there are other problems, too. Because their lives are fairly sedentary, animals in zoos develop some of the same diseases as overfed and underexercised humans. Obesity can be a serious problem, leading some zoos to pursue absurd fixes like a treadmill built for Maggie, a fat elephant in the Alaska Zoo.[13]

She refused to use it. Animals fed a suboptimal diet over many years can develop other nutritional problems, such as hyperparathyroidism in large cats and cholesterol granulomas in meerkats. Hsing-Hsing, the famous panda housed at the National Zoo in Washington, DC, had a favorite treat, which his keepers kindly provided nearly every day: the waist-expanding Starbucks blueberry muffin.

When individuals in captivity live to be senior citizens, unique welfare problems can emerge. One study, for example, found that aged zoo animals, despite appearing healthy, often suffer from chronic health conditions, such as painful osteoarthritis, that go undiagnosed by zoo veterinary staff.[14] On the other hand, over the past several years, as researchers and zoo staff have become better at recognizing the welfare challenges of geriatric animals, some zoos actually spend considerable energy and resources taking care of elderly or ill animals who cannot be displayed. Thus, it is not always true that animals living for a long time is good for a zoo's bottom line. Moreover, although some species and some individuals may have longer lifespans in captivity, there are many species and individuals who do not.

Girafficide, aka Zoothanasia

One of the management tools of zoos is so-called management euthanasia, or what Marc has termed "zoothanasia." Sometimes, of course, zoo animals are killed to relieve suffering. Animals become very old or terminally ill, and their pain cannot be adequately addressed with veterinary care. These account for some of the animals euthanized. But healthy animals are also killed, and for much more mercenary reasons. They are considered disposable when breeding programs are too successful or when a species is overrepresented.

The poster child for zoothanasia is Marius the giraffe. Marius lived at the Copenhagen Zoo until 2014, when he ceased to be genetically valuable. The zoo's stated reason for "euthanizing" the eighteen-month-old giraffe was that he was not going to be a good breeder, because his genes were already well represented in the giraffe breeding program. Several other zoos and sanctuaries offered to take Marius in, but Copenhagen Zoo proceeded with the killing. As further public spectacle—or, as the zoo saw it, as an educational opportunity par excellence—Marius was publicly dismembered and fed, in large pieces, to some of the zoo's carnivores. The most shocking thing is that Marius's case is not an isolated one. Animals considered unwanted, surplus, or genetically unfit are routinely killed at zoos around the country and around the world. It is part of normal zoo-management practices.[15] So too is what zoo managers call "breed and cull," where animals are allowed to reproduce but the offspring are killed.[16]

As we've seen throughout this book, one of the core ideologies of animal welfare science is that euthanasia of animals is acceptable. There is a strong preference for death over discomfort, and an unwillingness to accept death as a harm or even as a significant event in the lives of animals. This attitude detracts from our commitment to positive welfare, because we fail to acknowledge that the potential for future enjoyment might be valuable to animals. The acceptability of killing as a management strategy must be challenged. In April 2016 some good news came out of the Antwerp Zoo. It has introduced a no-kill policy for surplus animals. Healthy individuals will be allowed to live, even if the zoo doesn't have space for them.[17]

Captivity Is Harder on Some Than Others

As University of Guelph's Georgia Mason notes, the diverse species held in zoos vary "in the propensities for good captive health and welfare."[18] Some species tend to be healthier, longer-lived, and more fecund than their wild counterparts; others survive and breed less well in captivity, and seem to suffer more psychologically. Why?

One risk factor that is very clear, and for which there are excellent data, is home-range size. Mason analyzed home-range size in relation to risk of developing stereotypies in zoos in several species. She found that animals with large home-ranges are at high risk of developing welfare problems in captivity. Polar bears, for example, have an average home range of 80,000 square kilometers. The average size of a polar bear enclosure in a zoo is one-millionth of this. Polar bears in captivity have high rates of stereotypies such as repetitive pacing and route tracing. Lions have home ranges of 350–900 square kilometers; rates of stereotypies in zoos are 48 percent. Elephant home range is 1,500–3,700 square kilometers. Zoo managers and welfare researchers have concluded that optimal size for an elephant zoo enclosure is 1.24 acres, which is about 0.005 square kilometers. Almost two-thirds of elephants in captivity develop stereotypies.

Another risk factor is what a species of animal eats and how they have evolved to procure food, and how well their natural foraging behaviors can be replicated in captivity. According to Mason's research, "prey-chase" distances can predict stereotypic pacing in carnivores, as for wolves, who tend to chase prey over long distances, and are likely to develop stereotyped pacing. Another interesting set of research data on ruminant species suggests that browsers, who feed on woody shrubs, don't do as well in captivity as grazers, perhaps because the behavioral needs of grazers are easier to simulate. She also predicts that dietary specialists (who eat a very narrow range of food) may be at greater risk for poor welfare and stereotypies than generalists (who can eat a variety of foods and are flexible in foraging style).

One of the factors that seems to protect captive animals is phenotypic plasticity, or the ability to alter behavior to suit current conditions. For example, what are sometimes called "weed species" (such as deer and coyotes) seem to do well just about anywhere. Also, being sedentary and of limited range (like sloths and koalas) seems to "preadapt" a species to captivity, as does being gregarious (like flamingos), as long as they have friends with whom to interact. Sometimes the factors that allow animals to do well in captivity are mysterious to zoo managers and researchers. For example, ring-tailed lemurs seem to thrive in all sorts of captive environments, but nobody is quite sure why.

Understanding the factors that predispose animals to poor welfare in captivity can help zoos develop effective enrichment strategies. For example, mimicking prey-chase behaviors or foraging styles can allow animals to engage in at least a small range of food-related behaviors. It can also help zoo managers and animal advocates establish a list of animals that simply should not be held captive. This list must include large mammals such as elephants, dolphins, and whales, whose social organization and territories cover vast expanses in the wild. It must include large carnivores such as lions, polar bears, and wolves, whose prey-chase behavior cannot be mimicked in captivity.

Making Zoo Animals Happy

Zoos love to showcase charismatic polar bears, because they are moneymakers who attract a lot of visitors. But the bears don't fare well, and visitors often become concerned when they see the bears behaving strangely, such as rocking or pacing back and forth. Whether or not they understand what they are seeing, visitors find stereotypies disturbing to watch.

Just as for laboratory animals and farm animals, stereotypies are common in animals held in zoos and aquariums. One study, for example, found that polar bears spend an average of 11 percent of their day engaged in stereotypic behavior, and that stereotypy was correlated with higher fecal glucocorticoid concentrations, which mean higher levels of stress. Stereotypies are the physical manifestations of a pathology caused by confinement. In fact, this pathology has been given a name: zoochosis. The term was first used by Born Free Foundation's Bill Travers, who rightly identified the abnormal repetitive, obsessive behaviors of zoo animals as a form of psychosis.[19] Captivity literally drives animals mad. Animals in zoos can be seen pacing back and forth, tracing and retracing a particular route through their enclosure, plucking out all their feathers or pulling out all their hair, scratching or rubbing or licking themselves to the point of serious self-injury, biting the bars of their cages for hours on end, and engaging in what zoo managers euphemistically call "regurgitation and reingestion" (eating their own vomit).

Zoos can meet some of the needs of animals, but they surely cannot meet them all, and some animals' needs are harder to provide than others or are simply impossible to provide. When an animal's needs are not being adequately met, or when she faces "adverse conditions" such as frustration at not being able to perform a highly motivated behavior such as hunting or scent-marking, the frustrated behavior "spills over" into stereotypy. As we discussed in chapter 4, stereotypies are also thought to be caused by brain dysfunction brought on by stress-induced damage to the central nervous system. This stress-induced damage may be caused by the animal's current environment, or might be a result of earlier trauma. Either way, stereotypies are widely regarded as a serious welfare concern.

As we noted above, stereotypies are common in captive polar bears, though absent from the wild. Giving polar bears certain "enrichments" seems to help. For example, giving bears more dry land area and visual access beyond their enclosure have both been shown to reduce pacing.[20] Often, the more naturalistic and free-living a captive animal's environment, the fewer problems with stereotypies a zoo will see.

Enrichment studies emerged primarily in response to the prevalence of stereotyped behaviors. How can we make bored and frustrated animals stop engaging in abnormal behaviors? By giving them stuff to do. But enrichment is also part of another related trend in animal welfare science, specifically paying more attention to *positive* welfare states. "Welfare" is being redefined as not merely the absence of or ability to successfully cope with negative experiences, but the promotion of states of enjoyment and happiness. This is an important shift in perspective.

Enrichment emerged as a topic in animal welfare science in the early 1980s, in relation to how the quality of an animal's housing affects health and well-being—what is called "environmental enrichment." Enrichment research fell into some disrepute among scientists during the 1990s, over concerns that animal interests are not necessarily served by enrichments. Since then, attention to enrichment has grown. Zoos are starting to have enrichment programs and departments, and to keep people on staff as enrichment managers.

Some examples of zoo enrichments give a sense of what they try to accomplish. One of the most popular trends is to provide naturalistic zoo enclosures, which aim to give animals a tiny slice of home. A barren cage might be replaced with an "African savanna" full of tall grass and a few plastic "trees." Cement "rocks" might be added to a penguin exhibit. Monkeys might be given climbing structures, "vines" on which they can swing, and platforms that provide some visual diversity. Elephants might be provided a mud pit in which they can cool off. One study showed that giving animals the choice to go outside, whether or not they opted to take advantage of it, had a positive behavioral effect.[21] At the Brookfield Zoo in Chicago, the red panda exhibit is built around a giant magic tree. At semirandom intervals, steel cups emerge from "knotholes" in the tree. Sometimes the cups contain food and sometimes they don't, which the zoo's curator says is just like nature: sometimes food just appears. At the Woodland Park Zoo in Seattle, the orangutans are given cut branches or hay or banana leaves so they can build nests. For jaguars, food and spices are hidden inside logs, so that the animals will be stimulated to search their exhibit for the scents. Big Cat Rescue in Tampa rotates animals into a "vacation enclosure," which is much bigger and more interesting than the normal enclosures.[22]

At the Beijing Zoo, officials have undertaken an environmental-enrichment initiative that involves allowing animals access to privacy. As one zoo official said in a news interview, "Animals have the right to be seen as well as not to be seen . . . Beijing zoo is continuing to add animals while allowing them the freedom 'not to be seen.' The plan is so effective that some unknowing tourists have even been 'tricked' into complaining that the zoo has fewer animals than previously."[23] It may not seem like simply being watched would be a significant concern for animals, but research suggests that the mere presence of zoo visitors can be stressful for animals. For example, a recent study explored so-called "visitor effects" on little penguins and found that after exposure to visitors the birds showed increased aggression, vigilance, and avoidance behaviors such as huddling.[24]

Enrichments are designed to improve animal welfare, and sometimes even to make animals happy. But they have a secondary purpose, which is to increase the entertainment value of the animals on display. Bored animals are boring to watch. People want to see the otters and seals swimming; they want to see the lions moving around, not just lying on a rock or hiding behind a tree. So keepers try to get the animals busy doing things that are fun and interesting to watch.

According to a newspaper report, the Maryland Zoo has a full-time employee dedicated to enrichment, and one of this employee's tricks is to spray Calvin Klein Obsession perfume—which is made with a synthetic form of civetone, a civet cat pheromone—on surfaces in the big cats' enclosures, to encourage tracking and scent-marking behaviors.[25]

Dr. Hal Markowitz, a longtime advocate of enrichment, marvels, in his book *Enriching Animal Lives*, at how far we have come conceptually in understanding the reasons for enrichment and the ways to provide it. Yet he is at the same time dismayed at how little financial support is actually provided for enrichment. It remains "something of a nicety rather than a fundamental need in the eyes of many."[26] If we are to continue keeping animals in confinement, this needs to change. Making animals happier must be a top priority, and written into the budgets of zoo managers. Nevertheless, we need to remember that enrichment is just a Band-Aid solution. It serves, like the Valium given to SeaWorld's whales, to manage the symptoms. But it can't treat the underlying disease. Only freedom from captivity can really resolve the illness.

The Ethology of Freedom: Agency and Empowerment

We need to provide environments in which animals' behaviors and choices and actions matter, in which they can be agents of their own lives. In their essay "Giving Power to Animals," Hal Markowitz and Katherine Eckert present the rhetorical question: "How can we expect animals in our charge to be 'mentally happy' if nothing that they do matters?" For instance, animals are specialized in how they hunt for or gather food. "When we deprive them of the opportunity to exercise these abilities, we essentially rob them of their natural existence, their source of pride, their sense of well-being."[27] This is an important point, and challenges a common misperception about animals in zoos, namely that they have, in the words of one social media commentator, "won the lottery." These animals, our commentator continues, "get fed, taken care of, and get to lose their kill-to-survive mentality."[28]

There is ample evidence from a broad range of species, from rats to pigs to grizzly bears, that animals want challenges, they want to work. The truism about animals in zoos, that they have a cushy life because we give them everything they need, is both dangerous and wrong. "Making life easy" for captive animals doesn't do them any favors, and in fact deprives animals of the opportunity to enjoy challenges. Drs. Marek Špinka and Françoise Wemelsfelder have developed one of the key aspects of what we might call the "ethology of freedom": the idea that animals need a chance to express "agency."

Špinka and Wemelsfelder define agency as "the propensity of an animal to engage actively with the environment with the main purpose of gathering knowledge and enhancing its skills for future use," and as the "intrinsic tendency of animals to behave actively beyond the degree dictated by momentary needs, and to widen the range of competencies."[29] Animals need challenges to overcome because these provide animals with the chance to express agency, and to gain functional competence through such work. We know this is true for humans: a life without meaningful work or engaging challenges would be boring and unfulfilling. The same is true for other animals. Agency "forms an important condition for an animal's overall well-being and health."[30]

Challenges in the wild include evading predators, finding food, dealing with changing weather, illness, or social competition. The natural environment is complex and ever changing, and so animals are continually adapting and responding. Animals engaging with challenges build what Špinka and Wemelsfelder call "competence," an array of cognitive and behavioral tools and strategies that an animal uses to respond to novel challenges. For example, animals engage in associative learning to discern complex environmental cues (e.g., finding food); they engage in instrumental learning to overcome hindrances or resistance (food running away); they engage in inspective exploration of novel objects and situations (new food source?). In the wild, most animals live in an "open world"—an environment in which they can always expand their horizon of knowledge, and they are inquisitive. Environments are unpredictable and changing, which is why animals have evolved the capacity for behavioral flexibility. Animals interact with others both of their own kind and with other species within their home ecosystem, and engage in complex and dynamic social and observational learning and communicating.

Problem solving, exploration, and play are prominent facets of agency and competence. Problem solving is intrinsically rewarding for animals. We've known this for decades, but recent research into food and lab animals offers fresh data that could be usefully applied in the context of zoos. For instance, Jan Langbein and colleagues studied intrinsic motivation in dwarf goats. Goats were taught to discriminate between different shapes, and were given water as a reward for getting the problem right. When offered both "free" water and water attained through solving the problem, a third of the goats still performed the cognitive task.[31] Problem solving affects mood, too. Špinka and Wemelsfelder mention a study in which individual pigs were summoned to an automatic feeding system by distinct acoustic signals.[32] Pigs exposed to the cognitive challenge had better overall coping skills and showed fewer aberrant behaviors than pigs fed by conventional methods. In rats, combining a food reward with a cognitive task in a barren environment leads to overeating. Why? Because they are motivated to perform the task itself, and aren't just satisfying hunger. Exploration is also intrinsically rewarding, and animals like environments where they encounter novel objects or situations.

What animals need, say Špinka and Wemelsfelder, are "appropriate challenges." If a challenge is too severe, it will frighten an animal into a stress response; if it is too bland, it will fail to arouse attention. The notion of "eustress," or useful stress, captures the meaning of appropriate challenge: when an event is moderately challenging, stress has a positive overall effect on an animal.

Expressions of agency are rewarding for animals, regardless of whether there is any outcome. So, even if a lion's meal is going to be provided by a keeper, finding ways to allow the lion to use his problem-solving skills to acquire the meal will provide a measure of satisfaction. Agency also benefits welfare because of what Špinka and Wemelsfelder call the self-reinforcing nature of agency. Parallel to human studies of flow, there is evidence that intense engagement through focused attention and sustained concentration leads to feelings of happiness in animals. Animals may experience a sense of flow when they have opportunities to "initiate and maintain meaningful cycles of behavioral and cognitive effort."[33] Think of a monkey playfully swinging from vine to vine through a forest canopy.

Agency makes animals psychologically healthier and it also serves their long-term physical health. Engaging in problem solving, exploration, and play helps improve physical endurance and sensorimotor coordination, and increases the connectivity and density of neural connections, which, it has been suggested, could translate into greater behavioral plasticity and adaptability.[34] The consequence of suppressed agency is poor welfare. In simple, restrictive environments, animals cannot play, explore, or work at the levels they need to have a good life. Little or nothing is novel or interesting. They have reduced behavioral diversity. Behavior patterns become less versatile, often falling into stereotypies. Animals will sleep away their days, not because they are content, but because they are bored. Boredom, depression, and helplessness lead to heightened fear and anxiety, and a reduced ability to cope. Animals in impoverished environments heal more slowly, and so are likely to experience more prolonged periods of pain.

Ron Kagan, executive director and chief executive officer of the Detroit Zoo, and Jake Veasey have suggested the addition of two freedoms to the Five Freedoms, of particular relevance to zoo animals. These sixth and seventh freedoms pay particular attention to giving animals control. If animals are to be kept in captivity, we must seek to provide them the freedom to exert control over their quality of life, and freedom from boredom.[35]

Getting Beyond "Good Welfare"

In comparison to food- or laboratory-animal welfare, the literature on the welfare of confined "wild" animals is small. We really don't know what many of the ten-thousand-odd species being held in captivity need in order to cope with their situation. Zoo managers and welfare scientists are the first to admit this. Additional research will certainly continue to help zoo managers understand what they can do to optimize captivity. But all the welfare science in the world isn't going to set these animals free, and being free to live out their lives in their wild homes is what will most obviously improve their well-being.

As we've seen in previous chapters, welfare science has a peculiar vocabulary, a set of euphemisms and catchphrases that mask moral problems by covering them with a scientific sheen: "enriched," "humane," "necessity," "management euthanasia," "ex-situ breeding," "social grouping." Questions about what animals want and need are, as in other contexts, narrowly drawn. We ask elephants if they prefer plastic trees or cement trees; whether penguins like to swim in front of a viewing window or if they prefer having people watch from above; if lions prefer "hunting" a burlap sack hung from a tree or a piece of meat hidden in an old tire. These are self-serving research questions, and don't give the animals much of a choice.

Animal welfare science moves along an asymptote, getting closer and closer to "adequate welfare." But we'll never quite get there. Although research continues to flourish, it seems as if all we're doing is adding minutiae. The science of animal cognition and emotion has somehow not altered the direction of the trajectory. We need to step off this graph altogether and make a new one, with different goals to capture a different paradigm of thinking about animals.

Are Zoos Vital for Education or Conservation?

> Whatever thrill is to be derived from staring at a captive tiger is quickly dispelled by the animal's predicament. Awe gives way to abashment and then to a nearly inexpressible loneliness over being the only beast that does this to another. As such, any zoo, in whatever form, becomes not a demonstration of our prowess so much as a pathetically confused and protracted apology made to a series of wholly diminished and uninterested subjects.
>
> —Charles Siebert, "The Dark Side of Zootopia"[36]

Philosopher Paul Taylor proposed a basic ethical principle for our interactions with other animals: noninterference. Humans have a duty to "let wild creatures live out their lives in freedom."[37] Indeed, if what we really seek is the best possible well-being for animals, this rule should be our guide. As Marc observes, "Well-being is at its highest for individuals who are allowed to live their own kinds of natural lives in the wild (or what remains of the wild)."[38]

Animals have a strong interest in not being taken from the wild and kept in captivity, and we should respect this interest unless we have a very compelling reason not to. "Compelling reasons" for holding animals captive, according to the AZA, include education, conservation, and amusement. Do these "compelling reasons" really offer adequate justification for interfering with the freedom of animals? If the answer is no, then zoos ought to follow in the footsteps of the dodo and become extinct.

Zoos clearly amuse visitors, as the number of visitors to zoos each year suggests. But human amusement is not an adequate justification for exploiting animals and violating their freedoms. There are many ways for humans to amuse themselves that do not involve harming animals.

Despite arguments of zoo administrators to the contrary, there is very little evidence that zoos, in their current iteration, are educational. An average big-city zoo might have over two hundred attractions, and even if a family spent an entire day they would have only a few minutes to see each species of animal, particularly if they also want to stop and eat the zoo's overpriced hot dogs, popcorn, and cotton candy. According to one study of visitor behavior, the average amount of time people spent looking at each animal exhibit was thirty seconds.[39] Various studies have shown that few visitors actually read the educational placards placed by the animal exhibits. And when they do, they will find a list of factoids about the animals, but not a great deal of substance.

Jon Cohen, in "Zoo Futures," notes his surprise and dismay at the number of creationists on the staff at zoos, and the anti-intellectual, antibiology feeling he witnessed. It seems as if zoo staff assume that visitors don't want to be educated; they just want to be entertained. Zoo administrators don't want to drive away customers, so they keep their mouths shut about dangerous ideas like evolutionary theory.[40] And education about animals and their biology can't get very far without invoking evolution. Zoos invite a certain kind of "looking" at animals, but it is not a form of observation that really encourages us to consider who the animals are. "As institutions focused on the forced and sometimes frivolous display of wild lives," Stephen Kellert writes, zoos "can foster a sense of separation and even alienation."[41] Zoos may even take us beyond alienation, as Paul Waldau suggests, and imprison our imaginations, "locking us and our children into a mentality that requires coercion and domination of nonhumans."[42]

It is easy to imagine alternative educational platforms that would teach children and adults about wild animals and conservation and ecology without holding animals captive. For example, three-dimensional virtual zoos and well-produced nature documentaries could provide education without harming animals. The Sea Life London Aquarium has implemented such an idea; visitors can meet a polar bear and her cubs up close and personal, and can interact with them, and can watch a whale splashing in the water nearby. The animals are not real, but are actually computer images called "augmented reality." They allow visitors to learn about wild animals without keeping the animals in captivity.[43]

Indeed, a feeling of awe and respect for animals and nature is more likely to be fostered by allowing people to learn about animals in ways that don't involve captivity. Philosopher Dale Jamieson points out in his essay "Against Zoos" that zoo goers are much less knowl-edgeable about animals than backpackers, hunters, fishermen, and other people who say they are interested in animals and who actu-ally spend time observing animals in nature.[44] We realize that not everybody has easy access to wild or tame nature, but where possible a family day at the zoo might be better spent, educationally speaking, hiking or sitting in a park watching squirrels and birds.

At this point, there really are no compelling data to support the claim that zoos educate in any meaningful way, particularly in a way that might have beneficial trickle-down effects for the animals themselves. Admittedly, there also is no evidence that zoos educate in a bad way, except for the often-voiced concern that zoos send a message that keeping animals in cages for our entertainment and fun is appropriate.

With respect to the possible educational role of zoos, a 2014 study by the World Association of Zoos and Aquariums (WAZA) claims to offer proof of zoos' educational value, but the "proof" provided is a mixed bag. The study reported an increase "in respondents demonstrating some positive evidence of biodiversity understanding." But the increase was only slightly more than 5 percent of a very large sample, and failed to show how what people learn about biodiversity might contribute to any future interest in or commitment to conservation.[45] Without strong supportive data, the educational role of zoos does not provide adequate justification for locking up animals.

Now, what about conservation? Whether and to what extent zoos contribute to conservation efforts is more difficult to assess. Certainly, money that is raised by some zoos supports legitimate conservation efforts for wild animals. Yet it is very important to recognize that many zoos put forth a "conservation charade," professing a commitment to "saving wild animals" as a way to raise money and build public support without actually engaging in any meaningful conservation work. Despite the fact that many zoos are phasing out elephant exhibits, in March 2016 seventeen elephants from Swaziland were secretly flown to the United States, to be placed in the Dallas Zoo, Omaha's Henry Doorly Zoo, and Sedgwick County Zoo in Kansas. These three zoos were purportedly acquiring these animals "in the name of conservation."[46] There is an intriguing backstory behind the transport of these elephants. It turns out that Marc was contacted anonymously by someone on the ground in Swaziland, who alerted him to the fact that elephants were being loaded onto a plane bound for the United States. The elephants were being secretly transported to avoid a legal challenge. The basis of the challenge was that it is unethical and inhumane to take these animals from their homeland and imprison them in US zoos.[47]

Interestingly, conservation status in the wild seems to be a predictor of which animals do well in captivity. Animals who are vulnerable in the wild also seem to be vulnerable in captivity. So efforts to save these most vulnerable species should perhaps focus less on trying to keep them alive in zoos and more on trying to keep them safe and alive in their wild homes. This would mean preserving habitat in which animals would be free to roam. Furthermore, sacrificing individual animals on the altar of conservation is ethically problematic. As Ron Kagan wrote as recently as 2015, "Conserving a species through zoo efforts (also still lacking significant evaluation and evidence in many cases) may be a limited and hollow success if individual animals suffer in the process."[48] And such individual suffering seems part and parcel of current zoo conservation methods.

If zoos really are the key to survival for some species, perhaps they should become more conservation centers, less entertainment and spectacle, so that the emphasis in all aspects of care and management is on the well-being of individual animals, without compromises being made for the sake of profit and entertainment and visitor experience. These conservation centers would likely need to be closed to the public, or allow only very limited access, so that the focus remains always on the animals. And still, we will need to wrestle with the moral dilemma of trading individual freedom for species survival.

Zoo Futures: The Science of Animal Well-Being to the Rescue

Can existing zoos do better? Of course they can, but they can't do enough. What we need are not better zoos, but zoos that are so transformed as to be unrecognizable in comparison to today's models. A first step in transitioning away from the status quo is to acknowledge that many species of animals simply cannot be held in confinement without causing them significant harm. Zoos should stop exhibiting polar bears, lions, elephants, wolves, orcas, dolphins, giant pandas, and other animals whose lives in the wild involve large home ranges and a broad network of complex and changing social relationships.

As an example of this transitioning, a number of zoos have permanently closed their elephant exhibits, with no reported decrease in zoo attendance.[49] The Detroit Zoo, which was in the vanguard of this evolution, finally and permanently closed its elephant exhibit in 2005. Director Ron Kagan and others felt that the facility simply couldn't provide adequately for the social, emotional, and physical needs of the giant pachyderms. The last remaining elephants, Winky and Wanda, were moved to the Performing Animal Welfare Society's ARK Sanctuary in California, where they will live out their lives in relative comfort.[50]

Instead of investing in elephant treadmills and other elephant enrichments, the Detroit Zoo spent millions of dollars developing a snail exhibit. It acquired 115 Tahitian land snails of five different species, some of them endangered or threatened. Snails, it turns out, are fascinating, particularly when they can be observed in an environment that closely mimics their natural home. One can learn a lot about evolution and biodiversity from studying snails. In fact, the opportunities for learning are much greater than with the elephants, because the zoo is able to create a snail habitat in which the animals are free to act as normal snails, something the zoo could never achieve with elephants. The exhibit also teaches people about the important role of noncharismatic, "insignificant" animals in ecosystem stability and diversity.[51]

Zoos could make a contribution to conservation and education by helping visitors appreciate the beauty and value of noncharismatic animals such as amphibians. These animals require a different kind of looking than pacing polar bears or playful otters, because they don't seem to do that much. Amphibians often remain motionless for long periods. And they are often cryptic—they blend into their background and can only be spotted by a patient viewer, who is willing to spend far more than thirty seconds at an exhibit window. Rather than being boring to watch, these creatures, like the snails, have interesting adaptations and can provide a valuable learning opportunity about the role of often-unnoticed members of an ecosystem. A large number of amphibian species are threatened with extinction, and zoos could usefully direct conservation efforts to offset potential catastrophic losses.

A few days after killing the giraffe Marius, the Copenhagen Zoo killed two adult lions and two cubs to make room for a new male it had purchased. In a news report about the incident, a zoo official gave the following justification: "Because of the pride of lions' natural structure and behaviour, the zoo has had to euthanise the two old lions and two young lions who were not old enough to fend for themselves."[52] The muddled logic of welfarism is on perfect display here, as the zoo seeks to justify killing four healthy lions in order to make room for one healthy lion. Although we want animals to behave in "naturalistic" ways, sometimes this natural behavior is bothersome and gets in the way of our primary agenda: to create an exhibit that visitors will pay to enjoy. The killing of the two elderly and two young lions was "necessary." Welfarism in the zoo setting will all too often put the zoo first, and the animals second, making the claim that "zoos are for animals" continue to ring hollow.

The science of animal well-being seeks to flip the priorities: individual animals must come first. In all likelihood, animals would gather around the conference table and decide that the entire enterprise ought to be scrapped. Yet it may turn out that animals really *need* sanctuary, because their wild habitats are too full of humans, too fractured, too hot, too unstable, and too polluted—or nonexistent, as in the case of polar bears whose habitat is melting into the ocean. Zoos would then function as refugee camps. The notion of selling tickets so people can come through and gawk at the refugees would be recognized for what it is: an insult to the dignity of those who have lost everything.

CHAPTER 6

Captive and Companion

captivity late 14c., "imprisoned, enslaved," from Latin *captivus* "caught, taken prisoner," from *captus*, past participle of *capere* "to take, hold, seize" (see "capable"). As a noun from c.1400; an Old English noun was *hæftling*, from *hæft* "taken, seized."

JESSICA: GROWING UP IN A MENAGERIE

My childhood was filled with various pets, from the crazy Irish setter Jesse after whom I was named, to Snoopy the rat, the fourth-grade class pet whom I adopted and then decided to let free in our garage (much to the dismay of my father), to a cat named Estes. When my own daughter was young, she, too, wanted to fill her life with animals. Following family tradition, our house became a neighborhood zoo. We had three dogs, a cat, rats, guinea pigs, mice, a snake, a gecko, a tarantula, a salamander, and various fishes. All the kids wanted to play at our house, and all the parents knew that if their child had grown bored with a pet, I would be the sucker who would agree (after a long session of begging by my daughter) to take over the animal's care.

It was during the height of our pet-keeping craze that my research took me into a serious study of the literature on animal emotions and cognition. The more I read about what's really going on in the minds of other creatures, the more uncomfortable I started to feel about holding them captive in our home. I especially worried about the gecko, Lizzy. She was such a beautiful and exotic creature, and so alien. Yet the life we provided her was actually quite sad. She had fake vines, a few plastic plants, and a daily diet of crickets from a box—and that was her twenty-gallon glass world. She spent her days either standing motionless, or trying to claw her way through the clear wall of the tank. She lived about two and a half years, out of a natural lifespan of twenty-five.

I instituted a no-replacement policy, and as the various creatures finally passed on, our census dropped. I no longer felt comfortable keeping animals locked up in cages. I told my daughter, "Only dogs and cats."

MARC: WHAT'S A GOOD LIFE FOR AN OLD DOG?

Inuk was a very fit dog, getting regular long runs, as he was a mountain dog, and very healthy for thirteen-plus years. But he declined fairly rapidly due to a gastrointestinal problem, so the veterinarian to whom he went and really liked prescribed a large orange pill, as I remember it, that had to be shoved down his throat. There was no guarantee that the pill would work but it was worth a try. To say the least, Inuk hated the pills, and after having three a day for four days, he ran away when he knew the pills were coming no matter how softly I spoke to him. He'd cringe in the corner of his large outdoor run or scoot up the dirt road as best he could. No one seeing him would draw any conclusion other than he didn't want to take the pills. If Inuk were a human, and in many ways he was, there wouldn't have been a shred of doubt that the pills were not at all welcomed. Inuk also did not appear to get any better and clearly was telling me no more pills, please.

What to do? We considered different alternatives and then decided (without asking the veterinarian, but letting her know what we had decided to do) that because the pills weren't working and were causing him a good deal of unneeded and obvious emotional distress, Inuk should spend the last weeks of his life enjoying every single moment as much as possible. He loved ice cream, so that's what he got. Every day he got a frozen pint of ice cream and he worked on it for hours on end, tail wagging, ears up, and clearly enjoying every second of this special treat. And, most remarkably, after a few days, he had more energy, took longer walks up the road, played with some of his dog friends who lived up the road, and loved to snuggle once again.

So, am I happy with how Inuk spent the last few months of life? I am, indeed, even if he might have had a few more days on earth if he'd gotten the awful pills. Would I do something similar again? Yes, I would, without a doubt. Inuk had a great life and there was no reason he should have spent his last days agonizing over the big orange pills. That's what we decided was a good life for an old dog.

For many of us, questions about what animals want and need are part of our everyday life, although we may never use the language of welfare science or the Five Freedoms. Perhaps we live with a dog, and worry that he is lonely when we go to work for the day. Maybe we live with a cat, and debate with our cat-loving friends whether it is better to allow our cats outside, where they face some danger but have interesting things to do, or to keep them inside at all times, where they are safe from cars and coyotes, but where their behavioral needs for hunting and wandering may go unmet. Or maybe our child is begging us for a pet snake or guinea pig or bird, and we are trying to decide whether to say yes.

The difficulties of trying to know what an animal wants or needs arise poignantly when an animal becomes ill or aged, and we're faced with trying to determine whether they are still experiencing a "good life" or whether pain or suffering are dominating their experiential world, and hastening death through euthanasia might be a compassionate choice.[1] Most people want to do the best they can for their animals, but figuring out what "best" is can be tough. Indeed, it was the experience of trying to sort through these difficult questions with her dog Ody that led to Jessica's work on end-of-life care for companion animals, including her book *The Last Walk* and her involvement with the growing field of animal hospice and palliative care.

Moreover, although individual pet owners are generally well-intentioned, some common pet-keeping practices impose surprisingly heavy burdens on animals, such as keeping a lone goldfish in a small bowl or a single hamster or rat in a small cage. Furthermore, the various facets of the pet industry impose systemic welfare challenges on animals, such as problems that arise in large-scale breeding operations and at wholesalers who supply large pet-store chains with live animals, not to mention the millions of discarded or abandoned pets killed each year in shelters, pounds, and animal-control facilities.[2]

Those who advocate on behalf of companion animals would like to see more attention to addressing these broader welfare concerns. We are going to give a brief overview of welfare science as it relates to pets, and note a few industry-wide welfare issues. But most of our attention in this chapter will be on exploring freedom and preferences in relation to our most common animal companions—dogs and cats.

Animal Welfare Science and Pets

Given the number of people who live with companion animals, and the size of the pet industry, it might surprise you how little research has been directed at the welfare of animals kept as pets, either in the home environment or in pet stores and breeding facilities. Very few people even talk about the Five Freedoms for pets. Over the past decade, there has been a surge of research into the cognitive and emotional lives of dogs, and to a lesser extent cats, though how well this knowledge has translated into attitudes and behavior is open to debate.[3] For the hundreds of other species of animal sold at pet stores or auctions, much less is known about their physical and emotional needs and how captive conditions might or might not provide what they need.

Anthologies on animal welfare and animal ethics typically haven't included the category of pet or companion animal, perhaps because it was assumed that pets are pampered members of the family and that there are, therefore, no significant problems with their welfare. It isn't surprising, then, that welfare science in this realm is less well developed than in relation to farmed animals, research animals, and zoo animals. As dog experts Nicola Rooney and John Bradshaw remark, "Our understanding of the welfare of companion animals is both incomplete and fragmentary."[4] However, things are starting to change for the better. Some of the newer welfare books are beginning to include discussion of companion animals, and welfare researchers and veterinarians are starting to turn attention to the numerous welfare compromises experienced by pets. The welfare of dogs and cats in shelters has become an important topic of research and many people are also beginning to turn attention to the welfare of pets within the home environment.

Hundreds, perhaps thousands, of different kinds of animals are kept as pets, with only the human imagination as a limit. If a person gets it in his mind that he wants a slow loris for a pet, there is a very good chance that he'll find someone to sell a slow loris to him (legally or not), and he will be able to keep this slow loris in his home with essentially no oversight of any kind from a welfare committee or US Department of Agriculture (USDA) inspector or an undercover animal rights activist with a video camera. Even within much narrower categories, such as dog ownership, there are vast differences in how individual animals are treated and how well or poorly they fare in human company, even if they share a home with someone who claims to love them. Some dogs are treated with the utmost kindness, and others are treated horribly. Even dogs treated with kindness, however, often have important needs that go unmet. Although it is a good start, even love is not enough to ensure well-being for animals.

Values determine which aspects of welfare get attention, for example physical health, psychological well-being, or the ability to perform natural behaviors. And of course, in a broader sense, welfare science embodies a particular philosophy about what animals are for: they are here for us to use, for our own pleasure and benefit. As in the other venues we consider, the welfare discussion about pet animals rarely if ever questions the appropriateness of the pet-keeping enterprise. Yet, as we've said before, thinking carefully about the implications of what welfare science is learning about animal cognition and emotion, and even taking seriously the challenge of the Five Freedoms, can push us into a new way of thinking about pet animals. And, like millions of other animals, pets can benefit from a shift to the science of animal well-being, in which every individual matters.

The Pet Industry

The number of animals kept captive as pets is mind-boggling. In US households alone, there are an estimated 78 million dogs, 86 million cats, 96 million freshwater fishes, 9 million reptiles, and 12 million small animals.[5] These numbers have been steadily growing for the past four decades. Even in the economic downturn, the pet industry was one of the few that showed continued growth.

Unlike farming and laboratory research and especially zoos, where good welfare for animals lines up with productivity and quality, the same is not true within the pet industry. Profit is decoupled from welfare, which means that those in the business of selling animals have little economic motivation to care about what animals need. Animals are often sold cheaply, so it isn't worth the extra cost for a supplier or seller to improve conditions and reduce suffering and death, nor do people have to think carefully about purchasing an animal.

USDA oversight of animal wholesalers is spotty at best, and rules regulating the care of animals at these facilities are like the rules for agricultural animals: they aim to prevent the most egregious welfare violations, but give little to no attention to positive well-being. Ironically, the class of animals that should, intuitively, be given the greatest protections based on our level of empathy and interest, are actually given relatively little.

And unlike the other venues we've discussed so far, where those handling the animals are generally quite knowledgeable about the species under their care, adoptive pet owners often are total beginners and know next to nothing about the natural history or about the environmental or behavioral needs of a given animal. This is perhaps why the majority of fishes, amphibians, reptiles, and small mammals who are brought into the home live for only a year or two, if they are lucky.

Because pet owners don't always know a lot about what a prospective pet may need, they may rely on the pet store for advice, and may assume that what the pet store sells will provide an acceptable habitat for an animal. But pet stores often provide little guidance about how to care for a particular kind of animal and sell products that are inappropriate or inadequate for the animal's needs. For example, there are no standards for what size cage a given pet requires and some of the cages sold in stores are far too small. A hamster in a US biomedical research laboratory must be allocated a certain amount of space; at your local pet store, you can likely find several cages advertised for hamsters that would fail to meet these standards. To give another example, a recent trend in the pet industry is the nano-tank for pet fishes. These miniature tanks are meant to be so small and unobtrusive that you can put one on your desk (and some even come with a USB port so you can plug in your phone). You can purchase a tank that will provide a pet fish exactly 4 cups of water in which to live out his or her entire (short) life. This should be illegal.

Captive or Companion?

All of our companion animals are captive, even those who are our closest companions. But some pets are more captive than others, and some are more companion than others. In rough terms, the more fully able and willing an animal (a species or an individual) is to engage in a companionable relationship with humans, the more freedom, and thus happiness, the animal can potentially experience as a pet. Or, to phrase this another way, if we have to keep animals locked in cages because they would run away or cause harm to themselves or others if we didn't, they are more captive and will more likely experience the negative effects of confinement.

It is useful to distinguish between true companion species (for example, dogs and cats) and those animals (exotics, wild animals, reptiles) who are kept as pets but can't or don't often form companionable bonds with human species. We share with a number of other social mammals what is sometimes called the brain's "social network," the neurological, physiological, and behavioral systems that facilitate social bonding. This network includes hormones like oxytocin, which "reward" the brain for social bonding. Because these biological and behavioral systems are flexible, they can give rise to social bonding between species. Thus, a human interacting with her dog, and touching her dog's fur, will show physiological changes such as the release of oxytocin. Likewise, a dog being gently stroked by her owner also has increased levels of oxytocin. We experience pleasure from each other's company.[6] Animals who can form a bond with humans likely make better companion animals and have greater potential for a good life in human company.

Might there be some species of animal whom we should not keep as pets if we care about providing them with a "good life"? Or to phrase this another way, if we ask which animals can live in companionable relationships with humans and have a life with optimal welfare and a high degree of freedom, the answer, in our view, is going to be pretty narrow. Jessica takes this up in detail in her book *Run, Spot, Run*, but we would suggest that exotics and wild animals should simply not be kept as pets; reptiles and amphibians and birds likely cannot be provided with what they need by most pet owners; and fishes have more complex needs than most people realize—even the common goldfish. (As an interesting aside, the high court of India is discussing whether birds have a fundamental right to fly. "It is the fundamental right of the bird," wrote Justice M. R. Shah in a verdict on the caging of five hundred birds in a Gujarat market, "to live freely in the open sky."[7])

Let's focus on reptiles for a moment. The University of Tennessee's Gordon Burghardt, ethologist and world-renowned student of reptile behavior, argues that the best we can do for captive reptiles and amphibians is "controlled deprivation." In other words, these animals cannot be kept in captivity and still have a good quality of life.[8] Biologist and medical scientist Clifford Warwick, who has written about the morality of keeping reptiles as pets, describes "at least 30 captivity stress-related behaviors . . . regularly observable in most kept reptiles . . . such as hyperactivity and interaction with transparent boundaries, both of which involve persistent attempts to escape, and hypoactivity, which involves efforts to biologically 'shut down' from a poor environment."[9] Reading this reminded Jessica of Lizzy, her daughter's pet gecko, and Lizzy's constant clawing at the side of her glass tank, and made Jessica wonder whether she was wrong to subject Lizzy to such a sad existence.

Clifford Warwick, writing about exotic pets, suggests that biologists and species experts develop and make available to the public a list system for pets. This list would rank various species that one might consider acquiring as easy to extreme, in terms of being able to provide what the animal needs. The list would show which species of animal, "shown through clear and evidence-based assessment," might be suitable as a pet. Health and welfare needs of the animal, says Warwick, would be defined by the Five Freedoms principles, but would also include public health considerations such as likelihood of transmitting a zoonotic disease.

This list should not be limited to exotic pets, because even some very common pets like goldfish and hermit crabs are difficult to care for properly and suffer to some degree in any captive setting. We need to look at each species and each individual and ask whether captivity imposes undue burdens. It is likely that the only animals who can live companionably with humans, and have a meaningful life, are those who can live "free," without the confines of a cage.

The Least Captive Dog

Freedom, in the sense of "return to the wild," has no meaning for deeply domesticated animals like dogs. But freedom is still important to them and ideally we can find ways to live with dogs that allow them a large measure of physical and behavioral freedom. Of course, living in human society involves compromise, and indeed these compromises of freedom are integral to pet-keeping practices and are often necessary to keep our dogs safe. We neuter and spay, so that dogs can't mate. We keep them on leashes, restricting where they can go and with whom they can interact. We often scold them for barking. We put fences and doors around their home so they can't roam. Even dog training involves a careful and methodical process of constraining natural behaviors. A "good" dog is a constrained dog, and an unconstrained dog is often "out of control" (which can be a one-way ticket to the animal shelter). When piled one on top of another, these constraints can compromise welfare. As companions of dogs, we can do our best to balance necessary constraints against as full a measure of freedom as possible.

Barnard College's Alexandra Horowitz tries to imagine the "least captive dog" and suggests this is an ideal toward which the caretakers of dogs can aspire. Dogs need, at least some of the time, to be free from the leash, and free from restriction to a house. They need to be able to run free, and they need to be free to stop and smell the roses, particularly when those roses happen to have been peed on by another dog. They need to be able to sniff other dogs' butts and groins, no matter how squeamish their human companion might be about such behavior, because this is entirely appropriate and important in dog society.

Dogs also should have the freedom to approach only those dogs or people with whom they want to interact, and we should give them the freedom to love and form attachments with whom they choose. Dogs are selective about which humans they like, and we should respect their choices and not force them into interactions that make them uncomfortable (and that all too often result in bites, which are then blamed on the "mean" dog). Even the dog park, which many people think of as the epitome of dog freedom, can fail to be freedom-enhancing. Some dog owners will happily take their dog to the dog park, but will then micromanage the dog's behavior the entire time, yelling at the dog to "stop being rude" for trying to mount another dog, "stop sticking your nose into her butt," breaking up bouts of play, and advising the dog on which "friends" he should be spending time with. For some dogs, going to the dog park is not freedom-enhancing at all, but rather is stressful and overwhelming.

Who should be walking whom? According to Dr. Horowitz, the restrictions we place on dogs, for example, tethering them to human owners, are more serious that we might think. "Subjected to a person's decisions about everything from where to walk (down what routes, and when), whom to approach (which dogs and people), and what to investigate (which odors can be loitered on and which cannot), the dog has little independent choice."[10] We need to remember that dogs' noses are like our eyes—noses are their primary sensory window onto their world. Allowed to "walk themselves," dogs will spend about a third of their time sniffing.[11] Letting your dog follow her nose is a cheap and easy freedom-enhancer, and your dog will be grateful.

One of the areas of greatest welfare concern for dogs is "training," which we like to call teaching. This is also one of the areas in which a thorough understanding of dog behavior, cognition, and emotion by owners and trainers could make a world of difference. Although dominance-based training methods are perhaps less popular than they were two decades ago, too many people still try to get control over a dog by inflicting emotional and physical punishments.

The American Veterinary Society of Animal Behavior, in their position statement on the use of dominance theory in behavior modification, says "AVSAB is concerned with the recent re-emergence of dominance theory and forcing dogs and other animals into submission as a means of preventing and correcting behavior problems."[12] Many dog trainers misunderstand what dominance really means. While dominance is often taken to be a relationship based on force or aggression and submission, and is seen as inimical to freedom, it does not have to involve force or intimidation, nor does it have to be a freedom-inhibitor.[13]

The late Sophia Yin, veterinarian and dog-expert extraordinaire, highlighted the need for *leadership*, as opposed to dominance.[14] Dogs can maintain a sense of control and freedom, because they can choose to follow "commands" that are more like requests to do something than an exercise of coercion and force. Reward is fundamental to the organization of behavior, and rewards (food, positive social interactions) motivate learning and performance. Rewarding dogs with positive reinforcement such as a treat or praise or toss of the ball results in faster learning than punishment.

Teaching our dogs how to live successfully within our homes and our human environments is one of the most important things we can do for their well-being. Many pet dogs are confused about what they are supposed to be doing, and this causes them to be anxious and also results in them failing to "behave." And so-called behavioral problems are one of the primary reasons dogs wind up in shelters and why, once in the shelter system, many of them will be killed. Our companions could have substantially more freedom and happiness if they were more certain about what we expect of them. Better taught = more freedom.

The Cat Conundrum

Domestic cats are fascinating animals who can easily strain our patience and force us to make decisions that can deeply compromise their lives. For example, a recurring question is whether cats should be locked up indoors or allowed to roam free. Many a cat owner has struggled with the issue of feline freedom, and has made the conscious decision either to keep their cat confined inside or to allow the cat access to the outside. An e-mail from Marc's Italian friend Giulia Buttarelli nicely captures the dilemma: "I don't know if my cats would have preferred living shortly across the roads catching mouses or living for long time sleeping on my bed." Marc's friend Tom, who lives in the mountains outside Boulder, decided that his cat Wolfie was terribly unhappy inside and allowed Wolfie to go in and out during daylight hours. Wolfie lived to be eighteen and died a natural death. Another of Tom's cats, who was lacking front claws when she was rescued, has been enjoying the outside for two years now. However, a third cat who was given the same freedom disappeared after a couple of months. Tom has said that he laments the loss of his cat, but believes that giving them some outside freedom was worth the risk.

Conventional wisdom is that cats should remain indoors. This is standard advice from veterinarians, shelters, and even animal welfare organizations. The Humane Society of the United States (HSUS), for example, insists that all cats be kept inside. HSUS argues that (1) it is dangerous for cats outside and (2) that cats can be happy inside. But some ethologists, ethicists, and cat owners, not to mention cats themselves, are dissatisfied with this advice. Clare Palmer and Peter Sandøe, for example, argue that keeping cats inside serves *our* purposes, not the cats'. Looking simply at the needs and behaviors of cats themselves, it is very hard to argue that captivity indoors is best.[15] There are other arguments for confinement, such as the impact of cats on wildlife, but these are unrelated to cat welfare per se.

The way we live with cats has changed over the past few decades, perhaps more than any other form of pet keeping, even dog ownership. Nearly a billion cats are kept as pets in the United States alone. The majority of cats are now neutered, and because of availability of cat litter, cat food, and automatic feeders, many more cats live indoors, either full time or part time. They are no longer semi-free as in the past, living in barns or under porches. Cats are confined to even smaller spaces than dogs. Indeed, cats are often recommended as pets for apartment dwellers or those in small homes, because cats "need less space" than dogs and "don't need to run around and get exercise." How wrong this is! There is an epidemic of boredom among cats, evidenced in increasing numbers of behavioral and medical problems, including an epidemic of obesity. Cats are relinquished in large numbers to shelters, often as a result of behavioral issues such as urinating in the house, even though these issues often stem from frustrations with their living environment. Cats in shelters are far less likely to be rehomed and more likely to be killed than dogs.

HSUS is correct that being allowed outside can be dangerous for cats, but so too can being confined inside at all times. Depending on location, cats may have to contend with busy roads, with predators such as coyotes or cougars, with humans who have bad intentions, and with the possibility of injury or disease. But as with our children, we cannot protect them from all risk, and if we try to, for example, by excessive "helicopter parenting," we wind up hurting them in other ways. They cannot become independent, nor can they enjoy the satisfaction of taking risks and making their own choices. By keeping a cat inside, we protect her from certain dangers, but we also seriously compromise her freedom. Letting cats be outside may be what ethicist Bill Lynn calls a "sad good," a good that involves an element of moral risk and harm.[16]

Renowned animal experts Dennis Turner and Patrick Bateson agree that the United States veterinary policy on indoor cats is not necessarily best for welfare, and indeed, European veterinary groups disagree with the "keep cats inside" ideology. Being kept indoors may reduce certain types of risk, but it also increases the risk of other welfare concerns.

A leading candidate feature of the cat's surroundings that may influence disease is the cat's perception of control and predictability. The perception of or actual lack of ability to control their surroundings is perhaps the greatest stressor in the lives of captive animals. Captive animals have little to no control over who their social partners are; how much space they can put between themselves and other animals; what types, amounts and availability of food and water they can consume; where when and how they can eliminate; or the quality and quantity of environmental stimuli, including lights, noise, odours and temperatures.[17]

So what about a cat who has lived her entire life indoors, never stepping foot into the world beyond the door of the house? She doesn't appear particularly interested in going outside and doesn't sit by the door and meow. She may even show some fear and stay inside when the door is opened and she is given the option to go out. Is it "natural" for this cat to prefer the indoors? Or has the cat simply been conditioned to her constricted existence? In response to this kind of example, Palmer and Sandøe talk about "adapted preferences." In humans, adapted preferences are those produced through "'indoctrination, psychological manipulation, and the denial of autonomy' and in the context of unjust and oppressive background conditions."[18] Is a cat's desire to remain inside a "preference" that we impose on them for our convenience?

Many people are concerned about the costs to wildlife if cats are allowed outdoors. Cats like to hunt, and perhaps they even *need* to hunt. Their wild relatives are highly evolved predators. Yet if we are really concerned about wildlife, we might consider breeding and purchasing fewer cats. It is hypocritical to bemoan the wildlife deaths caused by outside cats and ignore the fact that we have created this problem in the first place, and have put cats in an impossible situation. Furthermore, we cannot blame cats while disregarding the far more serious human threats to wildlife, such as cars, highways, habitat destruction, and climate change. Ultimately, there is no solution to the cat conundrum that doesn't carry some costs, to individual cats, to cats as a collective, and to the birds and small creatures whom the millions of pet cats in the country would very much like to hunt and eat. The best solution is that we humans re-examine our pet-keeping practices and make more informed decisions about whether or not we really want to take a feline companion into our homes.

Preference Tests for Companion Animals

We assume too often that our animals think like us or want what we want and nowhere is the temptation stronger than in relation to the animals with whom we form close bonds. However, we need to pay much closer attention to what the animals themselves, as unique individuals, are telling us. One of the ways in which animal welfare science can offer huge benefits to pets and pet owners is in the application of preference testing. Much less research has been done on preferences of animals in the home environment than on animals in other venues, mainly because these tests emerged from within the animal welfare paradigm, and pets have mostly fallen outside the scope of welfare concern. But as Bob Dylan tells us, the times they are a changin'.

More and more researchers are turning attention to what companion species like and don't like within the home environment. For example, we are learning about acoustic preferences, and specifically about the responses of different species to music, which is very often a part of the human environments in which pets live. A study by University of Wisconsin professor Charles Snowdon and colleagues on music preferences in cats suggests that preferences for different frequency ranges and tempos vary from one species to another, based on species-specific communication systems.[19] Snowdon and his team composed special "cat music" to match cats' vocal range, which is about two octaves higher than human vocal range. The tempo of the cat music was set to match with pulse rates of purring or suckling cats. Cats were significantly more likely to respond positively to the cat-specific music than to human music. This kind of research is important because music is being used in some settings, such as shelters and animal hospices, to decrease stress and we need to know whether it actually does so.

Preference testing within the shelter setting is helping improve welfare. Shelter animals exist in an entirely different environment than pets living in a home, and so their needs are different. Shelters can be extremely stressful, but shelter managers are discovering various ways to improve the well-being of the animals who wind up there. Preference testing can help determine things like optimal size and composition of play groups for dogs; ideal acoustic conditions for cats; and visual aspects of shelter design, such as keeping cats visually isolated from dogs.

These days, preference testing isn't just for scientists, but also can be a valuable way for guardians to learn about and engage with their companion. Indeed, preference testing itself can become a form of enrichment. An all-time favorite game for many dogs is "what food do I like best?" which might involve presenting the dog with a variety of different food options (e.g., kibble, canned, homemade) and seeing which she goes for first, and whether she is willing to do some work to get to her favorite. Dogs also like to be challenged by puzzles. Research by Ragen McGowan and colleagues published in *Animal Cognition* shows that animals may experience positive feelings when they have successfully solved a problem, what researchers call a "Eureka Effect."[20] Dogs were more excited to receive a food reward after learning to complete a task than when given the food for "free." This is research you could try to replicate at home with your own dog.

Brian Hare and his colleagues in the Dognition lab at Duke University have been developing a series of interactive cognitive games that owners can play with their dogs.[21] The games are fun for the dogs, and help give owners perspective on how their dogs see the world. The "Dognition Assessment," as it is called, tests dogs in five areas of cognition: empathy, communication, cunning, memory, and reasoning. For example, one of the games will have you test whether your dog will follow your pointing or follow his memory, to find a treat hidden under a cup. You can enter the results of the various tests into an interactive online tool, and the Dognition team of experts will then tell you if your dog is a Maverick, a Socialite, an Einstein, or a Protodog. At the same time, the information you enter adds to Hare's database and helps increase knowledge into dog cognition.

Dogs clearly offer us a window into what preference tests can offer. They can also help us understand, in a personal way, that each animal is a unique individual. It is very hard to generalize about "the dog" and about what "dogs" want or need or like, because each dog is going to give a different answer. Everyone who has lived with, or been well acquainted with, more than one dog knows that they are as different as people, and this lesson can carry over into other venues in which we are thinking about who animals are and what they want.

Boredom and Frustration: Costs of Captivity

Pets, like other captive animals, can develop stereotypies, or what veterinarians often refer to as compulsive behaviors. In dogs, some of the more common stereotypies include tail chasing, chasing shadows or lights, and excessive licking or chewing on paws. Kenneled dogs will sometimes spin in circles or jump up on the kennel door, trying to get out. Cats sometimes compulsively lick their paws; parrots will pluck out all of their feathers or route trace. Rats and mice and hamsters will bar bite. An Italian study of stereotypies in pet rabbits, ferrets, and rodents found that just over a quarter of the animals in their survey engaged in stereotypies.[22] Often, particularly in dogs and cats, these abnormal behaviors are labeled as behavior problems and thus the fault of the animal, not their human captors. If the animal is lucky, their owner will seek the help of a veterinary behaviorist. If not, the animal may well wind up being punished for her suffering and often even may be relinquished to the shelter.

Humans can do wonders with training, and sometimes psychopharmaceuticals can be helpful. But it is much better to prevent animals from becoming so unhappy and stressed out in the first place. Stereotypies are only the most extreme form of psychological disorder. Most animals kept as pets suffer from boredom and from the frustration of not having enough control, enough agency, enough to do with the hours of the day. As Françoise Wemelsfelder explains, "To be able to create a meaningful life, the animal must be provided with materials that are biologically salient and enable it to fulfill its primary needs in an inventive, varying, and flexibly adaptive way."[23] And as we've seen in previous chapters, figuring out what these "biologically salient materials" are and how to provide them is a challenge that even welfare scientists haven't mastered.

Over the past decade, an increased awareness of companion-animal welfare has led to a growing literature on enrichments. Our homes, as comfortable as they are for us, aren't necessarily very comfortable or stimulating for cats and dogs. Their environment can be enriched by providing regular access to the outside and to social companionship. We can vary their diet, provide psychological "work" like puzzles and games, and offer physical outlets (e.g., things for cats to climb and scratch).[24]

Research into animal cognition and emotion has also led to reassessments of the capacities and needs of pet animals once assumed to be simple-minded and unfeeling. For example, there is growing recognition that fishes are cognitively complex, and need to be challenged with mental stimulation, the opportunity to socialize, and the chance to learn.[25] The days of the lone goldfish in a small bowl on the dresser really need to come to an end. Research suggests that captive fishes kept in enriched environments have improved neural plasticity and spatial learning skills, compared to fishes held in a traditional boring tank.[26] Enrichment of the environment can be achieved by altering periods of light and dark or adding novel elements to the tank, such as plants and rocks. Mental stimulation can be provided through behavioral training. For example, goldfish can be "clicker-trained" with a light to swim through a hoop and to come to the top of their tank for a food reward.

What Are We Doing with What We Know?

We already know a great deal about some animals kept as pets. There is a huge database on the preferences and needs of rats, mice, guinea pigs, ferrets, hamsters, and other so-called pocket pets. But the database exists within the realm of laboratory-animal welfare, and there has been very little movement of this information into the realm of pet keeping. There are no standards, as we noted above, for optimal cage size with respect to pet hamsters, rats, mice, or gerbils, nor are there standard recommendations about social groupings. And as we noted above, although we may have ample scientific data on the needs and preferences of animals in a certain venue, this data doesn't translate seamlessly into other venues, because the welfare challenges are different. Knowing, for example, what empathic and fun-loving rats in the laboratory setting prefer doesn't solve all the problems they may face as pets.

As we discussed in chapter 2, there is a kind of double-speak when it comes to animal emotions. The basic tenet of welfare science is that animals have feelings and that we should aim to minimize the bad feelings that they experience. At the same time, scientists are expected to remain skeptical about whether and what animals can feel and to be cautious of using language that might "anthropomorphize" animals and impute too much human-ness to them. This same double standard is also evident in veterinary medicine. Although many veterinarians talk freely with their clients about what a given animal may be feeling, there has been a tendency to shy away from using "emotion" language in the veterinary literature. When Jessica was working on a set of best-practice guidelines for veterinarians doing animal hospice, several of the veterinarians working on the document said they were uncomfortable using the word "emotion" to describe the experience of animals, and didn't want to describe animal feelings in terms like "happy" or "hopeful" or "depressed." According to their training, they argued, it is more scientifically rigorous to use nonsubjective labels when referring to animal behavior ("negative affect" instead of "sad" or "depressed," for example), especially when describing affective experiences. This is a perfect example of the knowledge-translation gap, because the scientists who study these "affective experiences" themselves use the language of emotions, and veterinary education isn't uniformly keeping pace with what ethologists are learning about cognition and emotion. Not only is it more scientifically accurate to use "subjective" terms, it is also more productive. Françoise Wemelsfelder and colleagues, for example, argue that using subjective terms ("fear," "frustration," "boredom") as direct interpretations of an animal's behavior, bypassing quantitative measures common to animal welfare science, is a more fruitful and ultimately more accurate approach. They call this "assessing the 'whole animal.'"[27]

It is exceedingly important that veterinarians speak clearly and accurately about the emotional and cognitive lives of animals, because vets are really the linchpin for animal welfare. They are the ones who can translate what we now know about who animals are and what they need into language and action that can be understood and applied by pet owners. If veterinarians fail to speak accurately about the feelings of animals, this in turn impedes pet owners' understanding of their animals, since veterinary communication is one of the most important ways we learn how to care for our companions.

The need for clear communication between animal and caregiver—and between caregiver and veterinarian—is nowhere more evident than at the end of life, when caregivers must make nuanced assessments of their animal's quality of life and where there is high potential for animal suffering. Jessica's book *The Last Walk* deals with this, and she is actively involved in the growing field of animal hospice and palliative care, one of the main goals of which is to help caregivers understand what their ill or elderly companions need and want. ("Caregiver" is the term favored by hospice veterinarians, to refer to pet owners.)

The most serious knowledge-translation gap—what is really a *knowledge* gap—is found among the pet-owning public. People often buy one type of animal or another on impulse, with absolutely no knowledge of the creature's behavior, biology, or social needs. They assume they'll just figure it out as they go. But this often translates into suffering for the animal, while the human learns through trial and error what a given species of animal needs (with "error" often meaning an early and uncomfortable death for the animal). This happens, most obviously, with exotic pets like geckos, bearded dragons, or turtles. But even with dogs and cats, the people who own and keep these animals often know almost nothing about their behavior, or worse, they have inaccurate beliefs, for example that we should understand dogs as miniature wolves. Many people simply choose to remain uneducated. This could deprive people of opportunities to provide the best possible life for their beloved companion.

The Animal Welfare Paradox

Animal welfare presents itself as an objective science that quantifies positive and negative physiological and psychological states in animals. It often claims to be value-neutral, but as we've seen in other chapters, welfarism actually embodies a strong ethical presumption, namely, that animals are here for us to use, manipulate, confine, and kill as we wish. To see this at play, consider that veterinarians can now become board-certified in animal welfare. The first principle of animal welfare, according to the American College of Animal Welfare, is that the use of animals for human purposes, for companionship, as well as food, clothing, work, education, and research, is ethical.[28] Principle 7 is that animals should be treated with respect and dignity. Within the welfarist world, there is no conflict between these principles. We can inflict harm on animals for profit and somehow call it "respect and dignity."

Welfarism appeals to our democratic spirit: it takes as a given that people—scientists, the public, animal advocates, and those with a financial stake such as zoo administrators or the farm-industry lobby or those making money in the pet industry—will have radically different opinions about the value of animals, what we owe them, if anything, and how much animal suffering matters. Welfarists will say, "Even if we disagree about whether animals have inherent value or rights or whether it is ethical to harm animals to satisfy our appetites or pleasure, we can agree that we should care about their welfare." But welfarism has failed to acknowledge the voices of the largest group of stakeholders: *the animals themselves*. The science of animal well-being, in contrast, makes its ethical commitments transparent and acknowledges that the scientific enterprise is, by necessity, infused with values. The stakeholders include individual humans and animals alike.

What do animals kept as pets want and need? They want what all the other animals we've written about in this book want: they want more freedom. Can animals in human company be free in any meaningful sense? We've suggested that for the vast majority of animals kept as pets, freedom in any meaningful sense is difficult if not impossible to provide, as is a good life of the sort we ourselves would want. For a few species, particularly dogs and cats, it seems possible that individuals can enjoy a large measure of freedom and can build a meaningful life within human homes and with human companionship. As it is now, though, many dogs and cats are not getting what they need, and a greater level of education and awareness among pet owners and veterinarians is essential. Knowledge about animal cognition and emotion, if uncoupled from the welfarist agenda, can help us learn what dogs and cats really need to be safe and happy.

Pet keeping is deeply enmeshed in our culture and is gaining in popularity. We are under no illusions that people will stop keeping pets, but we hope nonetheless that the trends will reverse as more people recognize that pet keeping is hard on the very animals we cherish and that the decision to bring an animal into the home must take into account who the animal is and what she needs.

For those people who have never been to a factory farm, stepped inside a research lab, or been behind the scenes at a zoo, the plight of animals within these venues likely remains unknown. However, companion animals can serve as a point of contact, helping us understand that the inner lives of animals matter. As people who have shared their life with a companion animal know, forming a close relationship with an animal can be a source of profound meaning and joy. Indeed, the animals of the world need a human population more inclined toward compassion and more knowledgeable about animal behavior, and one of the best ways for this compassion to take root and grow is through education and a deeper understanding of the human-animal bond, the subject of the growing field of anthrozoology. If we apply the concept of individual well-being to the animals we know and love best, perhaps we will then apply it more broadly to other animals as well.

CHAPTER 7

Born to Be Wild?

On July 1, 2015, the unhappy fate of an African lion named Cecil made international headlines. Cecil was killed by an American dentist who had traveled to Africa for a pricey trophy-hunting expedition. Cecil's demise had all the earmarks of a good story: the arrogance of a rich American, the underhanded way in which Cecil was lured from the safety of the game preserve he called home and where he was supposedly protected from humans, and the questions about whether the African hunting guides knew that what they were doing was illegal.

Six weeks after Cecil's story captured the public's attention and generated a huge amount of criticism, another high-profile killing occurred. Blaze, a grizzly bear living with her two young cubs in Yellowstone National Park, was killed by park workers after she attacked an off-trail hiker. Too young to survive alone, her two children were taken into custody. Instead of being rehabilitated and released into the wild, the cubs were sent to the Toledo Zoo, where they have bonded with an orphaned Kodiak brown bear whose mother was killed by hunters.[1]

Cecil and Blaze are high-profile examples of human intrusion into the life of wild animals. But their stories are hardly unique. There are, in fact, many ways in which human activities constrain or alter the normal species-typical behavior of animals in the wild and countless instances in which we can and ought to better respect the "wild needs" of the animals with whom we share the planet. As it turns out, even wild animals aren't truly free.

Welfare science is relatively undeveloped in relation to wild animals, because "welfare" is about responding to and, when possible, reducing the unpleasantness of harms that we deliberately impose on captive animals. Where wild animals have become captive—in zoos or homes—there is ample welfare literature, which we've already explored. But what about animals whom we do not routinely consider to be captive? We have a tremendous body of research into how human activities constrain the lives of wild animals and how we can mitigate some of these harms. Continued discussion about the ways in which we constrain the freedoms of wild animals is essential.

Although wild animals are not technically captive, they are nonetheless often deliberately manipulated by humans to achieve a certain outcome: a good balance of deer to hunters, for example, or of wolves to elk. Often these manipulations have the potential to cause harm. A distinction can be drawn between "management" and "conservation" of wildlife, with management focusing on controlling animals to meet human needs and conservation focusing on balancing the needs of some animals against the needs of other animals or ecosystems, for the sake of protecting endangered species or ecosystems. Nevertheless, in practice this distinction tends to get pretty fuzzy.

This chapter will focus on three different aspects of human interaction with wild animals. First, and briefly, we'll consider deliberate human intrusions, where the intention is to harm or kill wild animals. Second, we'll explore deliberate human intrusions that aim to help wild animals, either by manipulating ecosystems to maintain a certain balance of species or by researching particular species in order to better protect them from further harm and possible extinction. Finally, we'll look at some of the myriad ways in which humans unintentionally constrain the freedoms of wild animals by disrupting or polluting their homes and habitats.

Killing Wild Animals for Sport

The most obvious way in which humans impact the freedom of wild animals is through hunting and fishing. Although estimates vary and exact figures are elusive, somewhere between one hundred and two hundred million animals are killed by hunters every year in the United States alone, and the majority of these are killed not for food but simply for fun and for the "outdoor experience." The number of fishes killed by anglers each year is impossible to estimate, but easily surpasses the number of mammals killed by hunters. Although many areas only allow catch-and-release fishing (in order to keep fish stocks high), the practice doesn't mean that fishes suffer no ill effects and simply swim off into the sunset. It has been estimated that fishes who are thrown back into water after being hooked with bait suffer mortality rates varying from around 33 percent to more than 60 percent, depending on a variety of factors, including species of fish, type of hook, how the fish is handled, and how much time the fish spends out of water.[2] Fishes definitely get the wrong end of the hook.

It might seem that the only animals really impacted by hunting are those who die by a bullet or arrow if the hunter is a good shot, or who manage to escape the hunter but are wounded and die a lingering death. But even those animals who are not killed or injured, and perhaps not even the target of hunters, have their lives disrupted. Animals must deal with the vehicles, campfires, noise, sounds, odors, hunting dogs, and the stress of having humans present.

Not surprisingly, hunting has been shown to be extremely stressful to hunted animals, both those who are caught and those who are chased but get away. A well-known study by biologists Patrick Bateson and Elizabeth Bradshaw explored the physiological effects of hunting on red deer (*Cervus elaphus*).[3] When red deer were hunted by humans with dogs, the average distance each deer covered during the chase was nineteen kilometers (just under twelve miles). In addition to depletion of carbohydrate resources for powering muscles and disruption of muscle tissues, the hunted deer also had elevated levels of beta-endorphins and high concentrations of cortisol, suggesting extreme physiological and psychological stress. Indeed, the state of the deer after the hunt was worse than after a traumatic road accident. After this study was released in 1997, Britain's National Trust banned deer hunting with hounds on its land.

A study of wolves who have been heavily hunted also has shown that the individual animals experience physiological and psychological stress and that the effects of this stress extend well beyond the moments of terror during which the hunt takes place. Scientists examined hair samples from a population of Canadian wolves who are under heavy pressure from hunting and learned that the hunted wolves had elevated levels of progesterone, testosterone, and cortisol.[4] Researchers speculated that higher progesterone levels might reflect "increased reproductive effort and social disruption." Higher testosterone and cortisol may also reflect stress related to social instability, because wolves are highly social pack-living animals. One also wonders about the effects of stress on so-called Judas wolves, who are used as bait to lure other wolves to slaughter.

As ecologist Andrés Ordiz, who works out of the Norwegian University of Life Sciences, and his colleagues note in a study on bears, the "behavioral effects of living under predation risk may influence the dynamics of [human] prey species more than direct demographic effects."[5] In other words, living under the threat of being killed by humans affects populations of bears even more than the killing of a few bears. Hunting can force animals to alter what biologists call their activity budgets: they may, for example, increase their vigilance, leaving less time for foraging or other behaviors. Ordiz's team studied the movement patterns of brown bears in Scandinavia, before and after the start of hunting season. Bears were expected to increase their daytime activity as the winter approached and days got shorter. But instead, and because of the presence of hunters, they decreased daytime activity and moved more during the dark hours, disrupting their rest time. "Bears altered their movement pattern at a critical time of year," the researchers noted, "during hyperphagia, when they must store fat reserves before hibernation, which is critical for reproduction."[6]

Both the capture of wild animals for permanent confinement in zoos and homes and the killing of wild animals for sport are indefensible. There is no compelling human need for entertainments of this sort, and the loss of animal freedoms and lives is tragic.

Wildlife Management:
Deliberate Intrusions into Animals' Lives

Let's move on now to consider deliberate human intrusions that aim to "manage" wild animals, either by manipulating ecosystems to maintain a certain balance of species or by removing individuals or populations who come into conflict with humans. Wildlife-management issues include humanely "euthanizing" pest wildlife, controlling populations of "weed" species such as deer, relocating or killing "conflict" animals, reintroduction campaigns, welfare concerns related to the humaneness of lethal trapping and snaring, welfare considerations of capture and tagging practices, and so-called predator bounties, in which people get paid for animal bodies, skins, and ears.

In the United States, the organization most clearly tasked with the job of negotiating the balance between human and animal interests is Wildlife Services, which operates under the umbrella of Animal and Plant Health Inspection Services (APHIS) of the USDA. The stated mission of Wildlife Services is to resolve wildlife conflicts and to allow people and animals to coexist, but there seems to be relatively little emphasis on the "coexistence" part of this equation. According to APHIS, Wildlife Services killed an astounding 2.7 million wild animals from over three hundred different species in 2014, making it a pretty average year.[7] Although a small number of these deaths are accidental, most are deliberate and are efforts to deal with "problem" or "pest" animals. Animals become a nuisance to humans when they trespass on human settlements, steal food, or harm crops or livestock.

"Nuisance" and "Conflict" Animals

In Colorado, where we both live, one of the most common human-animal conflicts is the so-called conflict bear. Bears, not surprisingly, find the smell of human garbage pretty alluring and will sometimes wander down from the mountains into human settlements. Given their size and strength, bears are considered a potential danger to people. The story of Bear 317, which unfolded in Boulder during the summer of 2015, is an example of what typically happens to conflict animals.

Colorado has a two-strikes-and-you're-dead rule for problem bears who rummage through trash, break into cars or garages, wander into people's backyards, or walk the streets of town. Mother Bear 317 already had two strikes against her. And since they were still dependent upon her, so did her two cubs, 315 and 316. Bear 317 had been tranquilized and relocated, with her children, once already, and the three of them found their way back to town. It wasn't surprising, given the proximity of Boulder to the mountains, the extreme shortage of food available to bears that season, and the laziness of Boulder residents who failed to secure their garbage in bear-proof containers. Colorado Parks and Wildlife officer Larry Rogstad told a local reporter that he would be crushed when 317 had to be shot. "It's devastating. It's a sentient creature, a magnificent animal. It's an animal we spend our entire life admiring and trying to do good things for."[8] But, he says, he had no choice. When finally there was a clean shot, 317 was killed and her cubs relocated to the northeast Colorado-Wyoming border.

Contrast Rogstad's "It's necessary and I'm under orders" response to that of a similarly placed conservation officer in British Columbia. When Bryce Casavant was told to kill two eight-week-old black bear cubs on Vancouver Island, he refused. He just said no to the killing. And he was suspended from his job.[9] Like 317, the mother of the cubs was a "conflict" bear, who had twice broken into a mobile home to get at a freezer full of salmon. She "had to be destroyed." But killing the cubs was too much for Mr. Casavant. As of our writing, the two cubs are in holding. Thousands of people had signed a petition asking that the cubs be rehabilitated and released back into the wild.

Killing nuisance bears and other animals who come too close to human homes is not the only option. While it may be the only option that requires no sacrifice from humans, it is not the only possibility, and humans need to be willing to compromise and share the burden of our extreme overuse of nature's resources. Research suggests that stochasticity—that is, randomness—in the unpredictable availability of natural forage is what drives bears to seek human food, and that dependence upon "anthropogenic" food sources is reversible. Contrary to the assumption guiding Colorado Parks and Wildlife, bears who use urban areas don't necessarily continue to do so in the future. Rather, it depends on how much food is available in their wild habitat. In poor forage years bears use urban resources; in good forage years they don't. In other words, once a conflict bear, not always a conflict bear, and killing these animals is not the only response. Humans can take responsibility by using bear-proof trashcans and by otherwise not making food available for the animals.

Other nuisance animals who are routinely killed by wildlife officers or even by pest-control companies in the Boulder area include raccoons, bats, prairie dogs, foxes, pigeons, deer, coyotes, mice, and rattlesnakes.

CONTROLLING, MAIMING, AND KILLING
IN THE NAME OF MANAGEMENT

Some of the welfare concerns raised in wildlife management relate to how best to control entire populations of animals. Often, "control" is synonymous with "killing." To give an example, typical methods of controlling deer populations in the United States are rifles, muzzleloaders, and crossbows wielded either by wildlife service employees, hunters licensed by the agency, or recreational hunters. Some, however, argue that alternative, nonlethal methods such as reproductive control (e.g., contraceptive-laced bait) and strategically placed fencing can be just as effective and should be employed whenever possible.

An example of a specific welfare issue is the use of leghold traps (also called foothold traps) in management. These traps, which have a foot plate and spring-loaded "jaws" that clamp onto an animal's leg, are widely used in fur "harvesting" (particularly coyote, bobcat, raccoon, and otter) and in "removal" (killing) of pest animals. There has been considerable attention to whether the leghold traps could be refined so that the steel jaws don't break the animal's leg or cause undue pain. This is of particular concern in conservation research, where animals will be released after trapping. Jaws can be padded, and fitted with a device that emits a signal when the trap has closed, and leghold traps can be replaced by foot snares or nets. But as you'll notice, this entire welfare discussion assumes that trapping animals is acceptable and "humane" and even necessary. This sounds very much like welfare discussions about abattoir slaughter techniques that never question the necessity of slaughter itself.

A literature review of the welfare implications of leghold traps by the American Veterinary Medical Association notes several neglected issues.[10] First, "many commentators believe that restraining traps cause wild animals some degree of fear." The traps also cause immediate injury to many animals, including maiming, bone fractures, swelling, and hemorrhaging. Although most of the welfare literature has equated animal welfare with injury alone, being held and restrained is, in itself, highly distressing to many animals, and "limb restraint has been shown to cause more stress than cage enclosures for foxes and ferrets." Raccoons "seem to have particularly adverse reactions and . . . a high incidence of self-mutilation." The AVMA review goes on to note that many animals caught in a leghold trap die of exertion, exposure to weather, or predation. Furthermore, leghold traps capture many unintended species (up to 67 percent of captures).

An experimental annihilation of wolves in Canada is another real-life example of killing and maiming in the name of management. In 2014 the *Canadian Journal of Zoology* (CJZ) published a research article that presented the outcome of an experiment in mass killing. Eight hundred and ninety Canadian wolves in Alberta were slaughtered using aerial gunning, trapping, and poisoning with strychnine-laced bait. The strychnine also killed other animals who were not part of the study. Minimum "collateral damage" that was deemed acceptable by the researchers and the CJZ included 91 ravens, 36 coyotes, 31 foxes, 8 marten, 6 lynx, 4 weasels, and 4 fishers. Part of the methods section of this paper reads as follows:

> Wolf packs were located from a helicopter and one or more wolves per pack were captured using net-gunning techniques and fit with a VHF radio collar. Using a helicopter, we then subsequently attempted to lethally remove all remaining members of each pack through aerial-shooting throughout the winter . . . with the radio-collared wolves removed at the end of winter. We also established toxicant bait stations, using strychnine, to augment aerial shooting and to target wolves that could not be found or removed using aerial-shooting.[11]

Writing in response to the Canadian study, Gilbert Proulx and colleagues objected most strenuously to the use of strychnine bait, arguing that its use in an experimental setting is unethical. The effects of strychnine are well known, having been carefully studied in experimental animals: death by strychnine "causes frequent periods of tetanic seizures, occasional cessation of breathing, hyperthermia, extreme suffering, and death from exhaustion or asphyxiation, which typically occurs within 1–2 hours of ingestion." If the dose is low, death can take twenty-four hours or more. Not only do strychnine baits kill wolves, they also directly kill nontarget individuals mentioned above, as well as other animals by secondary poisoning, as when the carcass of a poisoned animal is consumed by a scavenger. The effects of "haphazard and indiscriminate" poisoning reverberate throughout ecosystems. As Proulx and colleagues note, "because fishers and wolverines have relatively low reproductive rates and large home ranges . . . the poisoning of just a few animals might jeopardize their populations."[12]

It's important also to note that this mass killing did not work. (Even if it had, it would not have been remotely justified.) As stated in the abstract of the research paper, "Although the wolf population reduction program appeared to stabilize the Little Smoky [caribou] population, it did not lead to population increase."[13] Cloaked as conservation biology, this egregious study raises serious questions about oversight and approval of lethal research involving wild animals. It is hard to imagine any other scientific investigation of a wild mammal being organized around the principle of mass killing. The methods used to exterminate the wolves are of the type used years ago and widely abandoned as unethical.

TRASH ANIMALS: HOW LABELING HARMS

How we label animals can be a freedom inhibitor for them, as is perhaps most obvious in the case of a "conflict" animal, "nuisance" animal, or "trash" animal. We label entire species in ways that have significant consequences. We'll see this below, with the Boulder-area prairie dogs who are considered a nuisance because they have the audacity to build colonies on land we think would make a nice location for an apartment complex or strip mall. And we'll also see this in situations where a species that is labeled endangered (spotted owl) is pitted against one who is not given this designation (barred owls).

Scholars, including journalist Fred Pearce and University of New South Wales professor Thom van Dooren, have recently begun giving attention to the issue of how we classify organisms, and particularly how we use the labels "invasive," "exotic," "native," and "indigenous." These labels are not scientific as much as rhetorical and political: the use of a label creates a set of attitudes. Invasive means out of place and dangerous (like an invading army). Being labeled invasive has huge implications for an animal or plant's protection. Animals who have been in an area for hundreds of years may still be considered invasives because they are nonindigenous. But as Pearce notes in *The New Wild*, many invasive animals have become part of the ecosystem, and the line between indigenous and invasive is often quite blurry. Pearce warns that transgressions of humane treatment are often justified by the labels "invasive" or "exotic." But invasive animals are still sentient beings.[14]

Another way in which labeling harms animals is when a species is demonized, particularly as a vector for disease. Bats, for example, have an image problem and are viewed as rabies-infested blood suckers. In a particularly virulent attack on bats, last year the government of New South Wales sanctioned the killing of whole colonies of flying foxes because they had been accused of spreading Hendra virus, which killed several domestic horses. Fruit bats are, indeed, the main host of the virus. But the only animal capable of actually transmitting the disease to horses is the domestic cat. Merlin Tuttle's 2015 book *The Secret Lives of Bats* explores how we might try to improve their image by reminding people that bats are important crop pollinators and consume vast numbers of unwanted insects like mosquitoes.

Moving Animals Here and There

People often assume that if humans and a certain animal or group of animals are in conflict, an appropriate and humane response is to move the animals. As we saw above, this often happens with "problem" urban bears. A bear caught rummaging through trashcans might be tranquilized and then moved back up into the mountains. Yet moving animals out of their home range can be problematic, and sometimes even fatal. An animal has to find food and shelter in an unfamiliar environment, and some must at the same time avoid being preyed upon or have to deal with human and nonhuman residents who are unhappy about the newcomers.

In wildlife management and conservation, "translocation" refers to the movement of an entire group of animals from one location to another.[15] Translocation is sometimes employed to increase the chances of survival for threatened species by, for example, increasing the animals' range or establishing new populations. Many translocations are spoken of as if they are undertaken for the sake of the animals, to "preserve species" or "promote conservation." In fact, these translocations are often mitigation efforts: humans plan a development project and need to move animal populations and communities out of the way. A study by San Diego Zoo researcher Jen Germano and her colleagues estimated that mitigation-driven translocations outnumber and receive more funding than scientifically motivated translocations, where actual conservation of species is the goal.[16] In some cases, the "mitigation" involves slaughter, as happens, for example, in many development projects in our home state of Colorado, where prairie dogs are routinely exterminated and entire colonies are gone as quickly as you can say "turn on the poison gas."

Even translocations meant to benefit only the animals themselves can involve serious trade-offs. A poignant example is the attempt to reintroduce wolves into areas where they used to flourish before being exterminated by humans. The last wolves in Yellowstone National Park, for example, were killed in 1926. In 1995 biologists proposed reintroducing wolves back into the park, primarily to control the burgeoning population of elk. Sixty-six adolescent Mackenzie Valley wolves from Canada were tranquilized and moved to Yellowstone. The new wolves were not the same subspecies that had once lived in Yellowstone, but a different one, identified by the Endangered Species Act as "experimental, non-essential," and thus legal to shoot and kill, if necessary, to protect livestock and manage population size.

The reintroduction was largely a success, and wolf populations are now established in the area. But the endeavor has been exceedingly complex, both ethically and practically. There were some violent battles between the wolves when they were establishing their new territories, with one pack hunting down and systematically mauling wolves from another. This is "natural" wolf behavior, but some of these territorial battles were created artificially by our interventions. Another major problem is that the reintroduced wolves prey on livestock, a very natural wolf behavior, and then they are killed by ranchers. This raises serious ethical questions about our responsibilities to translocated animals, and in what ways we are responsible for them in their new surroundings. For example, are we obliged to offer them protection? The case of the Yellowstone wolves highlights the classic dichotomy between an "animal rights" position, which focuses on the well-being of individual animals, and a "conservationist" perspective, which focuses on species, populations, and ecosystems.

Another example of what happens to animals when we move them around concerns the not-so-charismatic bamboo pit vipers found in Hong Kong. Researcher Anne Devan-Song placed transmitters on resident snakes and on snakes who had been relocated to a national park.[17] Most of the relocated snakes died within the first year of their relocation. As Devan-Song explained, "Snakes know their home range really well, so if they're dropped off someplace else, they take off and make all sorts of unusual movements that aren't typical of snakes. The more they move, the less time they spend eating, reproducing, and finding hiding places. Movement is a good indication of how well the animals were doing, and the relocated snakes moved a lot and didn't do well." Some of the relocated snakes were run over by cars, some were killed by other animals, and some appeared to have died from stress.

TRADE-OFFS IN CONSERVATION PRACTICE:
BLOOD ON THE TRACKS

As we've already seen, the distinction between wildlife management and wildlife conservation is rather blurry and can be rather bloody, and it is sometimes hard to determine whether a project is aimed to benefit humans, animals, or both. Now let's consider some conservation practices that are genuinely focused on benefiting animals, and which highlight some of the dilemmas and trade-offs that arise when we intrude into animals' lives, however well-meaning our intentions.

In a 2014 essay on the often dark history and current state of conservation programs, science writer Warren Cornwall highlights one of the common threads in conservation literature: the trade-offs that arise in our efforts to protect species, promote the freedom and well-being of individual animals, and balance both of these concerns against human needs.[18] Trade-offs between humans and animals are rarely decided in favor of the animals.

Some of the trade-offs are between two different species and involve killing individual animals in the name of conservation. For example, over the past decade, sea lions have been killed for eating too many salmon, and Arctic foxes have been killed because they eat Steller's eider ducks, as if eating salmon and ducks were a crime. The US government announced plans in 2014 to shoot 16,000 double-crested cormorants on East Sand Island off the coast of Oregon because they were consuming too many salmon to maintain viable populations of the fish. By the end of September 2015, 1,221 adult cormorants had been killed and more than five thousand nests had been destroyed.[19] And by May 2016, the East Island cormorant colony had collapsed.[20] Many people remain upset by this plan, not least the scientists who have been studying the cormorant population, particularly those who argue, based on scientific data, that the slaughter won't work. The blinds that scientists were using to study the birds, and which the birds have come to regard without fear, are the same blinds being used for the massacre. What a mean double-cross.

Owl Versus Owl

One of the most contentious conservation "trade-offs" of the past few years has been the decision to protect the northern spotted owl by killing barred owls.[21] The northern spotted owl lives in old-growth forests in the Northwest, and because of intense pressures from habitat loss due to logging is critically endangered. The barred owl is an East Coast species, but has been migrating west into the forests of the spotted owl. Barred owls are larger and more aggressive, and in forests where barred owls have settled, the spotted owls have been disappearing. Biologists are uncertain exactly how or even whether the barred owls are affecting the spotted owls, but in a desperate move to stop the process, the US Fish and Wildlife Service proposed in 2013 an experimental culling of barred owls to see if spotted owls might show some recovery.[22]

The experimental culling was highly controversial, and so the USFWS hired Clark University ethicist Bill Lynn to put together a series of stakeholder meetings to explore the ethics of the lethal experiment. In a long ethics brief written by Lynn, he raised serious concerns about experiments on wild animals. He noted that such experiments are subject to far less ethical scrutiny than experiments on lab animals, and lab animals have at least minimal federal protections while wild populations have none. Lynn noted, "Currently, there are no regulations in the United States that vigorously protect the well-being of individual wild animals in field experiments."[23] He also wrote that "many in the Barred Owl Stakeholder Group believed that the USFWS should take a strong leadership role in developing ethical guidelines for field experiments that explicitly take into account the well-being of individual wild animals." Nevertheless, Lynn concluded that it was acceptable to kill the barred owls as an experiment, as long as the killing was "humane." However, he balked at supporting a regionwide war on barred owls. Lynn called the killing of barred owls a "sad good." The Barred Owl Stakeholder Group didn't actually include any barred owls, but if it had we can bet that they would have strenuously objected to the decision. From the barred owls' point of view, the killing was what Marc called a "sad bad." On the other hand, if any spotted owls weighed in they may have welcomed the decision. They are suffering, too. Of course the best way to protect spotted owls is for humans to stop logging their forests. It is really not owl versus owl, but humans that set up the circumstances of competition.

Another kind of trade-off involving the sacrifice of certain animals for the sake of others is the practice of giving prey animals to carnivores so that they can "practice" hunting. This is sometimes considered necessary when a species is on the verge of extinction and there are plans to reintroduce into the wild members of the species raised in captivity. Plans by a group called Save China's Tigers, for example, reported that in order to reintroduce critically endangered captive South China tigers back to restored protected areas within their historic range in China, they have allowed the tigers to practice killing deer-like ungulates called blesbok.[24] Within the United States, measures to reintroduce endangered black-footed ferrets included breeding thousands of golden hamsters so that the captive-born ferrets could practice killing the hapless hamsters before being released into wild habitat. For many, this sort of "playing the numbers game" is unacceptable. Just because golden hamsters are plentiful and we can "make" as many as we need, doesn't mean they should be bred merely to be practice prey.

One thing to note about these trade-offs is that in nearly all of these instances, the conflict between species was spurred by the unrelenting spread of humans into wild areas. As habitat shrinks and pressures for food and space intensify, some animals will struggle more than others to survive. For example, Idaho is considering poisoning thousands of ravens because they are threatening the survival of greater sage grouse. But the reason ravens have such an advantage is that human encroachment on sage grouse habitat has left the animals vulnerable. Power lines and communication towers have provided perfect places from which ravens can spot sage grouse nests and steal the eggs. Although conservation officials may say that the only option is lethal control of one species over another, the real trade-off has already been made, and humans have forced all the animals into a losing position.

We are not going to suggest that trade-offs can be avoided, for in our overpopulated and challenging world they can't be. However, we are arguing that the barometer has to be adjusted so that the integrity of individual animal lives has a much stronger pull. There will be blood. That much is sure. But less blood and suffering are certainly better than more, and there are many ways to reduce the carnage.

RESEARCHER EFFECTS WHEN STUDYING WILD ANIMALS

It is essential to study animals in their natural habitats, and learning about them is critical to helping them thrive and survive. Yet one of the ironies of trying to learn about wildlife is that sometimes the way we study animals can harm them or alter their behavior. We must be aware of these possibilities. Research practices such as capturing and tagging animals can appear noninvasive, because they do not cause physical harm or obvious suffering, yet they can actually cause a great deal of stress and alter the behavior of the animals who are being studied, and this in turn can influence the integrity and utility of the data that are collected.

Patterns of finding food can be affected by human intrusions. One study found that foraging behavior of little penguins is influenced by their being fitted with a small device that measures the speed and depth of their dives. The attachments, even though they are tiny, nevertheless result in decreased foraging efficiency by the penguins. Another study found that the weight of radio collars influences dominance relationships in adult female meadow voles; when voles wear a collar that is greater than 10 percent of their live body mass, there is a significant loss of dominance. In yet another example, researchers discovered that simply placing a tag on the wing of ruddy ducks leads to decreased rates of courtship and more time sleeping and preening. Data gathered on mating patterns, activity rhythms, and maintenance behaviors of the tagged ducks would be misleading. Changes in behavior such as these are called the "instrument effect." Even the mere presence of humans can influence the behavior of animals. Magpies not habituated to human presence will spend time avoiding humans, taking time away from essential activities such as feeding.[25]

Study methods can be harmful in more direct ways, too, even though such harms are unintended. Radio collars, for example, can cause unintended physical damage, as graphically demonstrated in the case of Andy the polar bear, whose collar was so tight that he was slowly being suffocated and starved to death.

Even research practices that don't involve marking individuals with radio collars or tags can cause stress and fear. It has long been assumed that capturing or tranquilizing an animal for a brief time and then releasing the animal back to the wild will cause only temporary stress for the animal and is generally harmless. Yet this very common practice may have important consequences for animals, impacting their freedom in ways we hadn't anticipated. We don't really know the long-term effects of capture or tranquilizing, but some researchers worry that animals might experience prolonged negative effects. For example, Jay Mallonée and Paul Joslin reported that a female wolf named Tenino who was captured at one year of age and placed into captivity displayed symptoms consistent with post-traumatic stress disorder. Her method of capture involved being darted by helicopter and translocated twice, experiences that are highly stressful. She was placed into captivity because of her "participation in livestock depredation."[26] Mallonée and Joslin can't be sure whether it was Tenino's capture or her captivity which led to her emotional trauma, but they believe that the method of capture was primarily to blame.

COLLECTING WILD "SPECIMENS"

Methods of studying animals in the field are steeped in tradition and history, yet some of the habits of scientists studying wildlife could usefully be tossed in the dustbin. One of these is the collection of specimens. When Jessica was younger, she spent a week at an Audubon camp in the Southern Sierras in California. Each morning, the campers were invited to go along with the staff naturalist as he checked the live traps that he had set in various nooks and crannies surrounding the campsite. Usually the traps were empty and they went away disappointed, but one morning something could be seen crouching in one of the traps. The naturalist oh-so-carefully opened the trap and oh-so-gently picked up a tiny furred creature. The man's eyes got wide and a grin spread over his face. In hushed excitement, he told the group that this was a rare vole he'd never seen in this area. With a swift twist, he broke the vole's neck and stashed the body in a pouch, explaining what a great addition it was to the museum's collection.

Isn't it a strange impulse, to kill and collect the rare and beautiful and wild? Yet field biologists have a long tradition of collecting specimens, whether for the pleasure of adding to a personal or museum collection, or to offer proof to competitive colleagues that a given rare species exists in a given place.

"Voucher specimens," those that prove existence, generally come from populations that are small and declining and in danger of extinction, and thus the need for a voucher. But this makes them vulnerable even to the loss of a single individual. As Arizona State University professor Ben Minteer and colleagues argue, there are alternative methods of documentation that leave the specimen intact, including high-resolution photography and audio and visual recordings. This is particularly true in the case of species reappearances, where species thought to be extinct are rediscovered.[27] But even newly discovered species can be "collected" in ways that leave the animal intact. A recent news story described the collection of a new species of bee fly belonging to an extremely rare genus. The existence of *Marleyimyia xylocopae* was documented using a high-quality digital camera.[28]

Yet specimen collection continues. A "ridiculously gorgeous rare bird" called the mustached kingfisher was recently sighted and photographed in the forests of Guadalcanal in the Solomon Islands. In an essay in *Slate* magazine, Rachel E. Gross writes,

> It's hard to believe the mustached kingfisher is a real bird. First of all, it looks more like a Baltimore Orioles–themed stuffed toy than a bird. Second, it's only found on the remotest of mist-caressed islands, similar to the legendary Pokemon bird Articuno. Finally, it has long eluded human capture, with only three specimens ever before collected, all females. "Beautiful but very cryptic," is how birdlife.org describes it. "Very few sightings, and male plumage remains undescribed." Until now. Last week, a team led by Chris Filardi, director of Pacific Programs at the American Museum of Natural History's Center for Biodiversity and Conservation, identified and photographed the first-ever male mustached kingfisher.[29]

After catching this remarkable bird, Dr. Filardi wrote, "When I came upon the netted bird in the cool shadowy light of the forest I gasped aloud, 'Oh my god, the kingfisher.' One of the most poorly known birds in the world was there, in front of me, like a creature of myth come to life. We now have the first photos ever taken of the bird, as well as the first definitive recordings of its unmistakable call."[30]

All well and good, right? Not for the bird. It turns out it was a most unfortunate and final encounter for "a creature of myth come to life." The ridiculously gorgeous rare bird was "collected" as a specimen for additional study. Of course, "collected" means killed, a lame attempt to sanitize the totally unnecessary killing of this remarkable sentient being.

Killing in the name of conservation or in the name of education or in the name of whatever simply needs to stop. It is wrong and sets a horrific precedent for future research and for our children. Imagine what a youngster would think if he or she heard something like, "I met a rare and gorgeous bird today . . . and I killed him." Even if this handsome male were a member of a common species, there was no reason to kill him.

Animals Under Pressure

Let's move on now to what we might call indirect intrusions into the lives and freedoms of wild animals. These are situations in which humans are simply going about their business, not deliberately trying to harm animals, but managing to do so nonetheless.

One of the obvious ways in which human activity constrains the freedom of animals is through increasing destruction of habitat, with its various and long-term consequences: isolation of species, loss of migration routes, disruption of species-typical behavior patterns, loss of access to food and water resources, and crowding.

Consider, as one example, the effects of habitat loss on pumas (also known as cougars or mountain lions) in Southern California; these wild cats have to compete for space against twenty million humans. There has been a sharp decline in genetic diversity over the past eighty years because puma populations have been isolated by highways and human settlements.[31] Fragmented habitat is causing disruption in what geneticist Holly Ernest and her colleagues at University of California, Davis, call "genetic connectivity." Populations of pumas cannot interbreed because they are separated by dense human settlements and are thus caught in a genetic bottleneck. Genetic diversity is important because it provides a population with resilience, and some members of a group will be more resilient to disease or more adaptable to climate change than others. Genetic bottleneck is only one of many challenges faced by pumas in a fragmented and highly urbanized ecosystem. In Southern California, annual survival rates for pumas are low—only about half will survive in a given year—and most deaths are caused by humans. Pumas are hit by cars or shot by people with "depredation permits" that give them permission to shoot a puma who has killed a domestic animal.

QUIET, PLEASE!

Humans are noisy mammals. Our planes, trains, boats, and automobiles produce a constant din, as do our factories, our mining and fracking operations, and our cities and homes. The presence of human noise affects animals in a variety of ways, many of them harmful.

How human noise affects terrestrial species is of interest in current conservation studies. For instance, noise alters activity levels and echolocation calls of bats. Bats rely on auditory information to forage, and some human noise falls directly within their auditory range. One study of bat behavior in response to noise looked at a population of Brazilian free-tailed bats living near one of the largest natural gas extraction fields in the United States. Compressors at these extraction sites run continuously, all day every day. Bats living near a compressor showed a 70 percent reduction in activity level, compared to a population of bats not living near a compressor.[32] Less activity means less food gathered, less rest, less time spent mating, and lower chances of survival. Another interesting study focused on the hearing sensitivities of giant pandas, which are on the International Union for Conservation of Nature's Red List of Threatened Species. Because giant pandas are sensitive to ultrasonic vocalizations, they may be more disturbed by human noise than previously thought.[33]

Noise also affects birds. Boise State biologist Jesse Barber and his colleagues conducted an experiment to see how much the noise from a busy road would impact the behavior of migratory birds.[34] They created a phantom road, using a complex array of speakers, and positioned it near a resting place for migratory birds in a forested part of Idaho. When the speakers were piping in traffic noise, fewer birds stopped to rest during their migration. The noise altered the birds' assessment of the habitat as a suitable place to rest for a while. Noise also impacts reproduction, feeding, escape, and a whole range of other important behaviors in birds and mammals. A robin listens for worms in soil, and songbirds find mates through their calls; they aren't as successful at finding food or reproducing when they have to compete with urban soundscapes.

There's also "ocean commotion."[35] Noise pollution is a major problem for marine animals. Ambient noise levels in the ocean have been steadily rising over the past four decades, as shipping traffic has increased.[36] Additionally, noise from navy sonar, ships, pile drivers, and acoustic guns used to search for oil, gas, and other sources is stressing mammals, fishes, and invertebrates and leading to behavioral changes. Marine biologists believe that naval sonar leads to the death of many dolphins and whales each year. Cetaceans use sound to navigate, communicate, and find food. Increasing levels of ocean noise can dramatically affect their ability to engage in these behaviors. A study of harbor porpoises, for example, found that "low levels of high frequency components in vessel noise elicit strong, stereotyped behavioral responses," such as "porpoising."[37] Porpoising is just what it sounds like: jumping in and out of the water. It is energetically costly and may affect foraging and social behavior, including the potential abandonment of calves.

Although the behavioral effects of human noise on animals have been well-studied, relatively little is known about how noise pollution directly influences survival. A study by University of Exeter biologist Stephen Simpson and his colleagues examined the effect of motorboat noise on Ambon damselfish. Stressed fish were less responsive to simulated predatory attacks by their natural predator, the dusky dottyback. The damselfish stressed by the motorboat noise were more than twice as likely to be eaten by the dusky dottybacks. This study shows that human-caused noise directly affects the survival and reproductive fitness of these fishes.[38]

Freedom from Selfies

Wild animals often have their lives disrupted by tourists. In September 2014 a mob of tourists taking selfies and using cameras with flashes converged on a beach in Costa Rica to watch (and document for Facebook) olive ridley sea turtles laying their eggs—a once-a-year phenomenon. There were so many disruptive tourists that many of the turtles simply turned back to sea without nesting.[39] Development and tourism along beaches of the Atlantic coast is also creating problems. On Florida beaches, rare loggerhead and green sea turtles come ashore to lay eggs in the sand. Hatchling turtles emerge in the dark, when they will be less visible to predators, and use the dark sky to orient themselves toward the ocean. The lights from beachside hotels and condominiums, and the larger orange glow from cities, confuse the hatchlings, who often wander inland or along the sand instead of into the water, making them far more vulnerable to predation.

Even penguins at the bottom of the world aren't free from tourists and their selfies.[40] Tourism in Antarctica has increased dramatically over the past two decades, with almost forty thousand people visiting the icy continent during the 2013–14 tourist season. In addition to the tourists, Antarctica attracts a large number of researchers who come to study the behavior of penguins, seals, other Antarctic fauna, and our planet's deep climate history. Increasing tourism, along with climate change, may be placing penguins at risk for disease. Scientists believe that Antarctic species may have relatively weak immune systems because they have been isolated from some of the world's common pathogens, such as E. coli and avian pox. Human visitors bring pathogens with them, and this could have dire consequences for the penguin populations. A research team found evidence of several large penguin mortality events across the Antarctic since 1969, including the death of more than four hundred gentoo penguins in 2006 from avian pox.[41]

Droned Animals

A newly emerging issue for wildlife is the million or so recreational drones that are now buzzing around the skies. Sometimes birds collide with drones, or try to attack drones and get injured by the rotating blades. Drones also scare animals, disrupting time budgets and normal patterns of behavior. In one particularly telling study, scientists set out to see whether the presence of drones was disruptive to bears. A group of bears in northwest Minnesota had already been outfitted with GPS collars and small heart rate monitors, for a different study. A research team programmed autonomous quadcopters to fly over a group of bears in the forest. On the surface, the bears did not seem bothered by the drones—they didn't alter their behavior in noticeable ways, except that one bear stared at one of the drones for a while. But each time the drones flew above them, the bears' heart rates became markedly elevated, sometimes by as much as 400 percent above baseline measurements.[42] The effects of drones on wildlife are so severe, in fact, that the National Park Service in 2014 placed a temporary ban on drones in national parks while a long-term plan is formulated.

STARLINGS ON PROZAC

Human activities have some surprisingly subtle effects on the freedom of animals. Animals aren't free to behave normally because their brains, and as a consequence their behavior, are being altered by chemicals in the environment. Have you ever wondered, for example, whether birth control pills alter the reproductive biology and behavior of other animals? This is not an idle question, and has stimulated research centering on the effects of human pharmaceuticals on wildlife. Many pharmaceutical compounds, including birth control pills and antidepressants, are only partially metabolized by the human body. Some of the metabolites are excreted and wind up in wastewater systems and sewage sludge. People also dump unused drugs into the toilet or the trash without realizing the danger, further adding to the pharmaceutical load in the environment. Metabolites are taken up by invertebrates who are then eaten by birds or other animals.

Humans have developed a huge array of pharmaceuticals to alter the chemistry of our brains, to make us less anxious and less depressed. In the United States, about one in five adults is on some kind of mood- or mind-altering prescription drug. To study what the effects of psychiatric drugs might be on wild populations, researchers have been studying their effects in more controlled settings. For example, a team of researchers studied the effects of fluoxetine (Prozac) on the behavior of starlings.[43] The team captured twenty-four wild starlings, fitted them with transponders, and separated them into two groups housed separately in outdoor aviaries. Half of the birds were fed wax worms injected with fluoxetine; control birds were fed unspiked wax worms. The birds' feeding behavior was then monitored. The starlings in the fluoxetine group showed alterations in foraging behavior. Rather than eat several large meals with some snacking in between, the drugged birds just snacked. As the researchers explain, what may seem like a minor change can affect survival in the wild. Birds can change their body mass quickly. A large meal before night-fall helps ensure that they have enough insulation and energy to survive overnight without food. Eating excessive food during the day can also make a bird too heavy to escape predators.

Another research study looked at the impacts of sertraline (Zoloft) on perch and found that fish exposed to the drug ate considerably less than normal. Loss of appetite is a common side effect of the drug in humans.[44] Psychiatric drugs in the food chain also have what are called "knock-on effects." Damselflies exposed to oxazepam, a ben-zodiazepine used to treat anxiety and insomnia, showed no changes in behavior, although the drug accumulated in the tissue of the insects. But perch who ate the enhanced damselflies did show marked behavioral changes: the normally shy fish grew bolder, leaving their school more often and eating more quickly.[45]

Pharmaceuticals are part of a much bigger problem of chemical pollutants for animals, which we've known about for several decades thanks to Rachel Carson's prophetic 1962 book *Silent Spring.* Carson described the chemical pollution of ecosystems as a tragedy of loss— lost life and lost wilderness.

Chemicals are not the only human waste-product that gets into the bodies of wild animals. For a snapshot of some other subtle pol-lution-related constraints on the freedom of wild animals, simply take a look at the daily headlines. As we were writing this chapter, for example, we learned that there is so much plastic polluting the ocean that up to 90 percent of seabirds wind up ingesting it. Plastic in their stomach can lead to intestinal blockages, and toxins from the plastics leach into the birds' bodies. After twenty-nine sperm whales were beached in Germany, researchers performed necropsies and examined the contents of the whales' stomachs. (A necropsy is an autopsy performed on a nonhuman animal.) Among the items pulled out were a forty-two-foot-long fishing net and a twenty-seven-inch piece of automobile plastic, among other plastic litter.[46]

Compassionate Conservation
and the Science of Animal Well-Being

Scientific research that helps us understand the lives of wild animals is vitally important, as are intentional conservation efforts to protect them. As we've seen in other chapters, welfare science tries to mitigate our harmful actions, making death quicker or pain less severe. But it rarely questions the acceptability of harming animals for our purposes or the prioritizing of human needs and desires above those of animals. The basic tenor of our interactions with animals remains the same. The science of animal well-being challenges us to explore ways in which we can be less selfish and violent in our relationships with animals. It focuses attention on individual animals themselves and what they need and want, not just what we need and want. It seeks first and foremost to avoid harming animals, and encourages creative thinking about how to protect the integrity and freedom of animals.

And, we have a way to at least minimize the harm we cause other animals, including those who are living in natural or near-natural environments and are able to run, fly, or swim free. Marc and his Australian colleague Daniel Ramp have been very active in developing compassionate conservation, a rapidly growing international and interdisciplinary movement that focuses on the well-being of individual animals and advocates a strong a priori principle of "First, do no harm."[47] (The term "compassionate conservation" was first introduced by the Born Free Foundation.) Because of Marc's work in this area, he was invited to become a member of the United Nations Knowledge Network on Harmony with Nature, a program that recognizes the intrinsic value of nature and promotes a nonanthropocentric worldview, and embodies the same values as compassionate conservation. Compassionate conservation is not just "welfarism gone wild," but offers a completely new paradigm for thinking about ourselves in relation to wild animals. It is reshaping conservation ethics as we travel through the Anthropocene.

Compassionate conservation also recognizes that humans are

stakeholders, too, when there are conflicts between ourselves and other animals. In our challenging world, it is unrealistic to try to resolve conflicts without taking into account all of the stakeholders. This does not mean that human interests should necessarily trump nonhuman interests, but that we need to work on solutions that honor the fact that we have intruded, often inhumanely and abusively, into the lives of the animals. Examples of putting "First, do no harm" to work include using nonlethal paintball guns to deter pesky baboons in Cape Peninsula, South Africa, using Maremma sheepdogs to protect livestock in various locales when poisoning the predators changes the dynamics of ecosystems and reduces biodiversity, and using sheepdogs to protect little penguins from red foxes in Australia.

A wonderful project run by Wildlife SOS India stopped the horrific practice of using dancing bears for entertainment and money by showing the local people that they and the bears could benefit in other ways. Conservation biologists Kartick Satyanarayan and Geeta Seshamani note,

Wildlife SOS spearheaded a conservation success story in India by resolving the barbaric dancing bear practice in which sloth bear cubs were poached from the wild, brutally trained in inhumane ways and spent their short tragic lives at the end of a four foot rope dragged through towns and villages to earn for the indigent, nomadic community called the Kalandars. Wildlife SOS's initiative was to both rehabilitate the sloth bears held in captivity and the Kalandars themselves in alternative livelihoods. This in turn made a huge difference to the sloth bear population in the wild helping in its conservation. . . . Attempts at resolution involve creating safe spaces for the animals (rehabilitation centres), teaching people behaviours which do not lead to confrontation with the animals in question (awareness and education), but most importantly to inculcate a feeling for the animals in question, emphasizing adjustment and acceptance of the existence of wildlife close to our human habitations.[48]

Clearly, it is possible to work out conflicts in ways that balance the needs of all stakeholders.

Another project centering on human conflicts with leopards showed that a simple change of perception from "problem animals" to "problem location" resolved many human-leopard conflicts in human dominated landscapes without having to kill the leopards. The research, carried out by wildlife scientists and park managers in Sanjay Gandhi National Park in Mumbai, India, also called into question the effectiveness of capture and removal or translocation to new areas. The report concludes,

> Changing the focus from "problem animals," which has proved ineffective, to "problem locations" also changes the issues that need to be addressed and measures that need to be in place. It brings into consideration site-specific measures to reduce leopard visitations that may lead to negative interactions with people. This could include control of the domestic dog population, appropriate disposal of garbage and kitchen and medical waste, and better protection for livestock in safe corrals. It draws attention to providing better amenities for tribal and under-privileged people living in the area, including lighting, housing, sanitation (to reduce open defecation in forests, especially at night), and other public safety measures. Changing the focus from leopard to location also implies a change from reactive measures (such as capture after conflicts occur) to pro-active efforts (such as making safer surroundings to pre-empt attacks) that reduce negative interactions while enabling people and leopards to share the landscape.[49]

We can try to uphold the core values of compassionate conservation while seeking solutions to conservation and wildlife management problems. As an example, killing a "problem" animal like Boulder Bear 317 simply shouldn't be one of the options on the table. Rather than killing bears who rummage through trash, we ought to focus on education, such as getting people to use bear-proof garbage cans. Predation on livestock by carnivores can be managed through

fencing and fladry (flags strung along fencing), as opposed to shoot-
ing and poisoning. In each of these cases, the animal should not be
labeled as the problem; the problem is one of mismatched needs and
expectations between human and animal, and how to find a nonvio-
lent resolution to the mismatch.

It is incumbent upon researchers and nonresearchers alike to set
an example for future generations. We can look to a new generation
of researchers who take seriously the value commitments of compassionate conservation. We cannot and must not continue to spill the
blood of other species because it's convenient and supposedly gets
the job done. As humane educator Zoe Weil aptly notes, the world
becomes what you teach and so we must focus attention on teaching children the values of compassion, generosity, nonviolence, and
respect for others.[50]

To effectively mentor compassion in the Anthropocene we need
to learn to be compassionate ourselves. The science of animal well-
being, with its focus on individual animals, fosters a growing compassion for animals who depend on our goodwill.

The Anthropocene and the Sixth Freedom: Freedom from Extinction

We live in a time of unprecedented human influence on the natural
world if, indeed, there is anything we can still call "wild nature."
Our current epoch has been named the Anthropocene, a provoca-
tive term coined by Paul Crutzen and Eugene Stoermer in 2000 to
denote a geological age in which human activities are profoundly
and irreversibly influencing geological processes. We have committed all our Earthling neighbors to an ecologically precarious future in
which the Five Freedoms may increasingly be threatened by climatic
disruption and its ecological reverberations. Even if we stop violating
the freedoms of wild animals in obvious day to day respects—if we
stop killing problem wildlife, stop encroaching on habitat, and find
ways to protect their acoustic and visual environments—we will still
be impacting their freedoms well into the future.[51]

We might consider adding to the Five Freedoms a sixth, which we could call the "freedom from extinction." In our efforts to confront our environmental future, we ought to consider not just our own survival but also the survival of other animals. Even if animals don't completely die off, we are altering their habitat and behavior to such an extent that we've influenced the future course of their evolution. The sixth freedom is really the freedom of animals to live in habitats in which survival is possible—and not a mere eking out of a poor existence, but a thriving and joyous survival.

Humans also become less free in their interactions with wild animals. Relationships with animals are part of the cultural heritage of communities, and we are altering these centuries-old patterns. In the afterword to *Where Have All the Animals Gone?* by Dale Peterson, photographer Karl Ammann tells the story of one of the guides at a camp in Samburu in northeast Kenya. The guide, Julius Lesori, told them that when he was a boy he used to herd cattle and would see giraffes every day. His children have never seen a giraffe. If he wanted them to see one, he would have to take them into the Samburu Reserve.

The power of humans to influence wild-animal habitat and survival is nowhere more starkly on display than in what we might call the Radioactive Reserve. In 1986 a nuclear reactor exploded near Chernobyl, in what is now Ukraine. The explosion blanketed a large swath of land in radioactive fallout so severe that officials estimate it will be unsafe for human habitation for the next twenty thousand years. Nearly all people living near the explosion fled, and the so-called "zone of alienation" surrounding the remnants of the nuclear plant is now mostly uninhabited by people. With only a few stubborn and hardy human residents, the area has largely reverted to wild forest. Biologists have been startled to find that despite high radiation levels, animals are thriving. Wolves, red deer, wild boar, and elk are more abundant near Chernobyl than they are in nature reserves of similar size and habitat in Belarus and Russia.[52] Nor is Chernobyl unique. Flourishing wildlife has also been documented in the demilitarized zone between North and South Korea, and around the American naval base at Guantánamo Bay, where human encroachment on the landscape is at a minimum.[53] The case of Chernobyl suggests that the effects of nuclear radiation can be less damaging to animals than the effects of human population.

It is sad that we have to go to an area that suffered from the worst nuclear disaster in history to find animals who are truly free to live in peace and relative safety. Perhaps the take-home message for humans is that animals need some spaces or places that aren't overrun with people or their toxic waste. We need to be proactive in maintaining existing and creating new human-free zones for animals.

CHAPTER 8

Coexistence in the Anthropocene and Beyond
Compassion and Justice for All

> We need another and a wiser and perhaps a more mystical
> concept of animals. Remote from universal nature and living
> by complicated artifice, man in civilization surveys the creature
> through the glass of his knowledge and sees thereby a feather
> magnified and the whole image in distortion. We patronize
> them for their incompleteness, for their tragic fate for having
> taken form so far below ourselves. And therein do we err. For
> the animal shall not be measured by man. In a world older
> and more complete than ours, they move finished and complete,
> gifted with the extension of the senses we have lost or never
> attained, living by voices we shall never hear. They are not
> brethren, they are not underlings: they are other nations,
> caught with ourselves in the net of life and time, fellow
> prisoners of the splendour and travail of the earth.
>
> —Henry Beston, *The Outermost House:*
> *A Year of Life on the Great Beach of Cape Cod*

Humans engage in intimate and necessary relations with other animals, and in most of these interactions we hold the power. But power is not a license for domination or abuse. Trying to imagine a world without human-animal interaction is both absurd and sad, especially since we evolved together. But can we imagine and perhaps create a world in which our interactions with animals are more respectful of their own needs and interests? We think the answer to this is a resounding yes! However, working toward such a world will require that we stop using science and human-centered arrogance as tools of violence against other animals. We need to move beyond welfarism.

Where Is Welfare Science Going?
The Welfarist Vortex

Animal welfare science is going strong and has firmly developed into an internationally recognized field of research. But where exactly is it headed? On the one hand, there have been some positive changes on behalf of animals. In March 2016 China released its first set of guidelines for the more humane treatment of laboratory animals, and the United States Congress passed reforms to the Toxic Substances Control Act, one of which requires that the Environmental Protection Agency reduce and replace animal testing for chemical safety where scientifically reliable alternatives are available. The editorial board of the *New York Times* called for the Pentagon to put an end to the use of live animals in combat-medic training. The Buenos Aires Zoo is closing after 140 years, citing as its reason that keeping wild animals in captivity is degrading. Iran banned the use of wild animals in circuses, and at the time of this writing 42 airline companies have adopted bans on trophy-animal shipments on their carriers.[1] We recognize that these are positive moves; however, the science of animal well-being will require more thoroughgoing changes.

And as time goes on, we are accumulating more precise data about the wants and needs of animals. Donald Broom and Andrew Fraser, two of the world's leading welfare researchers, write, "Our knowledge of . . . welfare indicators has improved rapidly over the years as people with backgrounds in zoology, physiology, animal production and veterinary medicine have *investigated the effects of difficult conditions on animals*."[2] Welfare concepts have been refined and methods of assessment have been developed, expanded, condensed. We have a good list of things that "challenge" animals: exposure to pathogens, tissue damage, attack or threat of attack, social competition, excessive stimulation, lack of stimulation, absence of key stimuli (e.g., "a teat for a young mammal"), and inability to control one's environment.[3]

In addition to the data, the Five Freedoms seem to be evolving conceptually. For example, David Mellor, of the Animal Welfare

Science and Bioethics Centre at Massey University in New Zealand, has suggested a shift in terminology to the "Five Domains." The domains model addresses certain weaknesses of the Five Freedoms and offers, according to Mellor, a more scientifically up-to-date method for assessing harms to animals. One of the key problems with the Five Freedoms is that the language "freedom from" in four of the five statements implies that the elimination of certain experiences (hunger, fear, pain) is possible. In fact, as we all know, these affective experiences are part and parcel of life and serve, biologically, to motivate an animal to engage in behaviors essential to survival. Mellor claims that the goal of welfare science should not be to eliminate these experiences, but rather to balance them against positive affective experiences.[4]

None of this amounts to a substantial evolution in the fundamental moral or scientific tenets and tenor of welfare science. Mellor acknowledges that the welfarist paradigm allows for negative welfare states, but he encourages a kind of reweighting of the scales so that the suffering we impose is tempered by tossing animals a few extra "positive welfare state" crumbs. He admits that animals will still experience pain and suffering, but wants to give them as much comfort, pleasure, and control as possible and reduce the intensity of negative states to "tolerable" levels, within the context of using them as we wish. We are still caught in the "welfarist vortex," and are simply accumulating bigger and bigger piles of data about how exactly we are harming animals and what they are experiencing within the various "challenging" situations we impose upon them.

While some may argue we are being too critical or not paying attention to the number of changes that have been made to improve the lives of other animals, welfare science continues to favor our interests over those of other animals and to patronize animals by acknowledging only their most superficial needs. There are new welfarist data— lots of new data—and this information is filling in what we know about how best to "humanely" slaughter, trap, confine, and constrain. But the value commitments of the welfarist enterprise are so strongly biased in favor of human self-interest that our treatment of animals under this regime will never move beyond exploitation and violence.

We may try hard to give animals a better life, but a better life is not necessarily a good life.

The moral commitments (or in our minds, the *immoral* commitments) of welfarism have remained constant: we are still the purveyors of pain and suffering. In what kind of world do we live when an entire research program is focused on how best to harm animals, and how to salve the conscience of those who might have reservations about the violence?

The Shield of Welfare Science

Hebrew University's Dr. Yuval Noah Harari, the author of the landmark book *Sapiens*, wrote an opinion essay for the *Guardian* in 2015 calling industrial farming the greatest crime in history. "The scientific study of animals," he writes, "has played a dismal role in this tragedy. The scientific community has used its growing knowledge of animals mainly to manipulate their lives more efficiently in the service of human industry."[5] Harari has captured the essence of why welfare can never be good enough. Animal welfare science operates in the service of a variety of industries, and while in this role it can and will never do more than reinforce the status quo. It will never challenge the brutal exploitation of animals in farming or in laboratory research, zoos, pet stores, or conservation-research programs. Indeed, as Harari suggests, science hasn't just been silent about our violent treatment of animals; it has lent its support and expertise to the endeavor.

Worst of all, welfare science has woven a cloak of objectivity around abusive practices. Broom and Fraser write, for example, that, "the assessment of welfare can be carried out in an objective way that is independent of any moral considerations."[6] Like Harry Potter's cloak of invisibility, the objectivity of welfare science is meant to shield those wearing it from moral examination. But the status quo that welfare science perpetuates *is* a set of value assumptions, including the assumption that the feelings of animals don't really matter all that much, and even if they do matter a little, their interests can be trumped when doing so serves our interests.

Science has been put to work to make our manipulations of animals more efficient, more productive, and more profitable. It has been a partner in crime with industries that use and abuse animals, and has been employed to substantiate and scientize and ethically neutralize crimes against animals. But this is not an inevitable role for science. Science has the potential to help animals and to heal our fractured relationship with them. Indeed, as the science of animal cognition and emotion continues to advance, it may well be that the weaknesses of welfarism will become more apparent and the basic inconsistencies will be laid bare. The more we know about the inner lives of animals, the more incongruous animal welfare science in the service of industry becomes.

Science, Ethics, and Advocacy

The basic insights of animal welfare science are profoundly important. The first of these is that animals have subjective experiences. The second is that animals not only experience negative feelings like pain and fear and frustration, but also experience pleasure, happiness, excitement, and other positive feelings. Following on these, the final insight is that behavior offers a clear window into animal feelings.

Behavior is, indeed, a good window through which to see and know animals. But it can be a very tiny welfarist window, in a house we design, build, and manage for our own ends. Or, it can be a much bigger window, one through which we can peer but didn't build, the dimensions of which are unknown. If we looked inside an abattoir or peered into an orca tank at SeaWorld, we would see a vast collection of "welfare" concerns. But the abattoir and the orca tank need to be seen from a much larger vantage point. We shouldn't be looking *in* the abattoir and the orca tank and tinkering with the conditions we find, but looking *at* them, taking full measure of what these places mean for animals. The essence of the ethology of freedom is that behavior is a window onto what animals really want and need—to be free to live their own lives, to be free from the suffering and exploitation to which we subject them—but only if we are looking the right way: straight into the eyes of the animals themselves.

In contrast to welfare science, the science of well-being uses what we are learning about cognition and emotion to benefit individual animals, continually seeking to enhance their freedom to live their own lives in peace and safety. To the three basic scientific insights of welfare science, the science of well-being adds the essential ethical corollary that the feelings of individual animals matter. In contrast to welfarism, a science of well-being acknowledges up front that science and values are intertwined and that our assessments of what individual animals need are scientific *and* ethical. Indeed, values come first and inform the kinds of scientific questions we are open to asking and the kinds of answers we are willing to discover. Welfarism is a cage that traps human perception, one that also confines our sense of empathy for other beings. We need to open the doors of the cage.

There will always be trade-offs in what humans need and what animals need. Humans inevitably interact with and use other animals, and we are not advocating a hands-off approach to animals and nature, although that might not be a bad idea in a human-dominated world. But a great number of things that we currently do to animals are simply wrong and need to stop: the unnecessary slaughter of animals for food and fur, the use of animals in invasive research, the confinement of animals for human entertainment, and our excessive encroachments on wildlife. The threshold for taking away an animal's freedom or denying any or all of the Five Freedoms is, at present, extraordinarily and offensively low. The bar must be raised.

As we've emphasized throughout this book, the central question motivating animal welfare science is "What do animals want and need?" This question has remained the focus of welfarism over the past five decades. Do we know enough to answer this question? Absolutely. We know enough, right now, to know that animals want to be free from human exploitation, free from captivity, and free from the sufferings we impose on them. This is not to say that further scientific research into animals' hearts and minds isn't important, for it is. The more we know, the more mindfully we can interact with other animals, as long as we can break out of the welfarist cage and focus more objectively on what *they* want and need.

What we must do now is to close the knowledge-translation gap.

We must apply what we know about emotion and cognition, and follow through on the moral implications of the science we currently have at hand. Cognitive ethology, the study of animal minds, needs to take a "practical turn," putting what we know about animals into the service of animals themselves. Scientists can be tools of industry, or they can be advocates for animals in ways that really serve the animals. We would like to see more scientists move away from being advocates for welfarism and become more positive advocates for the animals themselves. While some scientists claim that scientists should not be advocates, they forget that arguing *for* the use of animals is advocacy that works against animals. A few years ago, Marc gave a talk in Sydney, Australia, where he argued that it was wrong to kill kangaroos for sport, fun, and food. At the end of this talk, a scientist working for the kangaroo-meat industry criticized Marc for being an advocate. He said that science is supposed to be objective and scientists should not be advocates. Marc responded that he and his critic were both advocates. Marc advocated for the kangaroos, whereas his critic advocated against them. The room got very quiet.

The best hope for closing the knowledge-translation gap lies with future scientists and with all of our children, because they have not yet been inoculated against compassion for animals. One can do "good science" and still feel for animals, and indeed, we've already seen that compassion and concern for animals can produce better science. Once this knowledge becomes integrated, business as usual will look very different.

By encouraging schools and parents to include humane education we can hope to raise children who both understand that animals have feelings and, more importantly, translate this into their daily lives and choices. Marc has been writing a lot on the notion of "rewilding education," retuning our relationship with the earth, and getting youngsters off their butts and out into nature.[7] A recent report has shown that prisoners in a maximum-security facility in the United States are guaranteed two hours of outdoor time every day, whereas 50 percent of youngsters worldwide spend less than an hour outside each day.[8] Not only will our children benefit from preemptive education, but so too will future generations as we negotiate the challenging and frustrating path through the Anthropocene.

What research into animal cognition and emotion continues to demonstrate is just how intertwined we are, evolutionarily. Human exceptionalism, the idea that we are of a different sort altogether, and thus (in our own self-serving logic) have a right to do as we please, is scientifically unsupportable. Writing about the 2015 discovery of fossils from an early human relative called *Homo naledi*, renowned primatologist Frans de Waal wrote, "We are trying way too hard to deny that we are modified apes. The discovery of these fossils is a major paleontological breakthrough. Why not seize this moment to overcome our anthropocentrism and recognize the fuzziness of the distinctions within our extended family? We are one rich collection of mosaics, not only genetically and anatomically, but also mentally."[9]

Fostering Freedoms

As we were in the early stages of writing this book, Marc received an e-mail from Jennifer Miller, who was working at a reintroduction center for previously captive parrots in Costa Rica. Jennifer told him the story of a great green macaw who had escaped from the center. The fate of the parrot became a source of argument among the center's staff. Jennifer's feeling was that they should not try to recapture the animal and should just let him be free. Others strongly disagreed, feeling that it was their obligation to find him and bring him back because he would likely perish on his own in the wild. This story is a wonderful example of how freedom for animals means different things for different people, and how freedom can conflict with other values.

We decided to ask some colleagues to share their thoughts about what freedom means for animals. Here are some of their responses:

Michael Tobias (award-winning author and filmmaker):
"We have no idea what freedom means. But we can certainly appreciate what the lack of freedom means."

Sarah Bexell (Institute for Human-Animal Connection, University of Denver): "Self-determination . . . including the choice of where to roam, fly, swim, choice of friends,

choice of activities, choice of food, choice of mates, choice
of home/nest, and even poor choices that end their lives,
but at least death came in the midst of freedom."

Jo-Anne McArthur (filmmaker for the video *The Ghosts in Our
Machine* and author of *We Animals* and *Captive*): "To be free
from bodily and psychological exploitation by humans . . .
to be respected by humans and not objectified."

George Schaller (world-renowned conservation biologist):
"An intriguing question. I just returned yesterday from
eastern Tibet in search of nonhuman animals. An animal in
the wild is free to spend much of its time in search of food
or starve, competing for status and mates, and remaining
alert to avoid becoming prey. A captive animal is fed well,
its social life, if any, confined to cell mates, and, secure from
danger, its existence is blunted and banal, its evolutionary
force spent, placing it among the living dead."

Hope Ferdowsian (physician and bioethicist): "The same
as for humans. Freedom to meet our basic physical needs,
whatever those might be by species and individual—
including freedom of movement (bodily liberty); safe
and secure from harm from humans (bodily integrity—
and this should include freedom from harm to the mind);
freedom to love and bond with whom we wish; respect
for our choices, and freedom from humiliation and
intentional shaming."

This is a sampling of what freedom means to people who have
worked in diverse sectors of the human-animal interface. But the
story of the macaw reminds us that we need also, and especially,
to think about what freedom means to animals. What did freedom
mean to the escaped bird? To be free to fly but possibly not survive
long, or to delay freedom of flight until better equipped to survive
longer? Maybe he gave us his answer by escaping.

Transitioning from Welfare to Well-Being:
The Adjacent Possible

A recent issue of the *Atlantic Monthly* featured as its Big Question "Which contemporary habits will be most unthinkable 100 years from now?" One of the responses was, "Eating animals for their protein."[10] It is indeed possible to imagine a future in which people will look back at how animals were treated in the early twenty-first century and shudder with horror. "They were barbarians," they may well say about us. "How could they possibly ignore animal sentience and suffering?" They might say this about all of the venues of animal use about which we have written.

Steven Johnson, who has studied and written about the history of innovation, explores the notion of what he calls the adjacent possible.[11] The adjacent possible, writes Johnson, "is a kind of shadow future, hovering on the edges of the present state of things, a map of all the ways in which the present can reinvent itself." The past and present prepare us for any number of futures. Depending on what groundwork has been laid and what ideas are floating around, certain new thoughts become thinkable. As Johnson suggests, "The strange and beautiful truth about the adjacent possible is that its boundaries grow as you explore them. Each new combination opens up the possibility of other new combinations."[12]

The pieces are here right now for a major paradigm shift in how we think about and interact with other animals. Indeed, they have been here for quite a while, but few are bold enough to say "enough is enough." A future is possible in which humans and other animals coexist peacefully, where nonviolence is the norm rather than the exception, and where exploiting animals will be viewed as morally offensive. Welfarism raises the ante by acknowledging that animals have feelings and that these feelings matter. But in continuing to favor human interests above the interests of individual animals, it doesn't go nearly far enough.

Enhancing the freedoms and well-being of individual animals, and championing the peaceful coexistence and harmony of animals and people, opens the door to a new adjacent possible. The Anthropocene—the Age of Humanity—may well evolve into the Compassionocene. Building on the momentum of increased global concern for the well-being of individual animals, we must work toward a future of greater compassion, freedom, and justice for all. This is the right thing to do.

Acknowledgments

We thank Alexis Rizzuto for her dedication to this project and her excellent guidance from the beginning to the end of writing this book. Thanks also to Dale Peterson for introducing us to Alexis, Susan Lumenello and Melissa Dobson for their fine copyediting, and Nicholas DiSabatino for helping with publicity. Thomas D. Mangelsen gave us the beautiful cover for which we are ever grateful.

Marc: I've had wonderful opportunities to work with numerous people around the world who are deeply passionate about protecting other animals and their homes. I've acknowledged their contributions in other books, and most recently I've had the pleasure of benefiting from the insights of Jane Goodall, Canadian parliament senator Wilfred Moore, Sarah Bexell, Thomas Mangelsen, Todd Wilkinson, Camilla Fox, Daniel Ramp, Arian Wallach, Louise Boronyak, Chris Draper, Dror Ben-Ami, Liv Baker, Jennifer Miller, Brooks Fahy, Joanna Lambert, Dave Crawford, Hope Ferdowsian, Wendy Keefover, Brandon Keim, Giulia Buttarelli, Catherine Doyle, and Dale Peterson, who have always been willing to discuss many of the issues with which we deal and for offering "on the ground" suggestions. Valerie Belt, Betty Moss, and Peter Fisher constantly send me notices about pertinent essays, for which I am most appreciative, and Bill Simmons is always good for asking hard questions on long bike rides.

Jessica: Thanks to Bella and Maya, for making sure I knew when the UPS guy was coming to the door and when it was time to take a break for snacks, and for providing a million daily reasons to care about the well-being of animals. To Chris, for his enduring support and his willingness to listen to horror stories. To Roger and Alexandra, for talking over ideas, week after week. Finally, thank you to Marc for sharing your enthusiasm, your wit, your expansive knowledge and experience, and your dark chocolate. It has been a treasure working with you.

Notes

Shortened citations appear
in full in the bibliography.

CHAPTER 1   FREEDOM, COMPASSION,
AND COEXISTENCE IN THE HUMAN AGE

1.   Christina M. Colvin and Lori Marino, "Signs of Intelligent Life: Pigs Possess
Complex Ethological Traits Similar to Dogs and Chimpanzees," *Natural History*,
October 2015, http://www.naturalhistorymag.com/features/122899/signs-of-intelli
gent-life; "The 10 Smartest Animals," NBCNews.com, http://www.nbcnews.com
/id/24628983/ns/technology_and_science-science/t/smartest-animals; Carolynn L.
Smith and Sarah L. Zielinski, "The Startling Intelligence of the Common Chicken,"
Scientific American, February 1, 2014, http://www.scientificamerican.com/article
/the-startling-intelligence-of-the-common-chicken; Melissa Hogenboom, "Rats Will
Save Their Friends from Drowning," BBCNews, May 14, 2015, http://www.bbc.com
/earth/story/20150514-rats-save-mates-from-drowning; "Something to Crow About:
New Caledonian Crows Show Strong Evidence of Social Learning," *Science Daily*, Au-
gust 26, 2015, http://www.sciencedaily.com/releases/2015/08/150826113817.htm; Ellie
Zolfagharifard, "Elephants Get Post-Traumatic Stress Too: Calves Orphaned by the
Killing of Their Parents Are Haunted by Grief Decades Later," *Daily Mail*, November 6, 2013, http://www.dailymail.co.uk/sciencetech/article-2488384/Elephants-post
-traumatic-stress-Calves-orphaned-killing-parents-haunted-grief-decades-later
.html; Marc Bekoff, "Fish Determine Social Status Using Advanced Cognitive Skills,"
Psychology Today, March 17, 2016, https://www.psychologytoday.com/blog/animal
-emotions/201603/fish-determine-social-status-using-advanced-cognitive-skills.

CHAPTER 2   CAN SCIENCE SAVE ANIMALS?

1.   The phrase "a life worth living" is slowly creeping into the welfare literature. David
Mellor, for example, argues that we have a responsibility to give animals "a life
worth living," but that it is not realistic to provide animals with "a good life." See
Mellor, "Updating Animal Welfare Thinking," 21. Temple Grandin also uses the
phrase "a life worth living," to describe what we offer animals in industrial food-
production systems. See, for example, Jennifer Demeritt, "See No Evil: Grandin

Designs Around Animals' Needs," Mediaplanet, http://www.impactingourfuture
.com/advocacy/see-no-evil-designing-around-animals-needs.

2. Bekoff, *Why Dogs Hump and Bees Get Depressed*.

3. Helen Proctor and her colleagues conducted a systematic review of the scientific
    literature on sentience. Using a list of 174 sentience-related keywords, they reviewed
    more than twenty-five hundred articles on animal sentience and found a sharp and
    steady upward trend in the number of articles published on sentience from 1990 to
    2011. "Evidence of animal sentience is everywhere," they concluded. Proctor and
    her colleagues discovered "a greater tendency for studies to assume the existence of
    negative states and emotions in animals, such as pain and suffering, than positive
    ones like joy and pleasure." This is consistent with the historical trend of scientists
    being more willing to accept that animals can suffer than to afford them positive
    affective experiences. An interactive website called the Sentience Mosaic, launched
    by World Animal Protection, continually updates the huge trove of data on animal
    sentience. "Discover Sentience Mosaic," http://www.worldanimalprotection.org/our
    -work/education-animal-welfare/discover-sentience-mosaic.

4. American Veterinary Medical Association, "The Veterinarian's Role in Animal Wel-
    fare," September 2011, http://www.acaw.org/uploads/AVMA-VetsRoleInAW-20116.pdf.

5. European Parliament resolution of May 5, 2010, on evaluation and assessment of
    the Animal Welfare Action Plan 2006–2010 (2009/2202(INI)) 2011/C 81 E/05, http://
    eur-lex.europa.eu/legal-content/EN/TXT/?uri=CELEX:52010IP0130.

6. Cambridge Declaration on Consciousness, July 7, 2012, http://fcmconference.org
    /img/CambridgeDeclarationOnConsciousness.pdf.

7. Jones, "Science, Sentience, and Animal Welfare," 18.

8. See, for example, Andersen and Kuhn, *The Sustainability Secret*; Clive, *The Animal
    Trade*.

9. Edward Berdoe, describing Claude Bernard's attitude toward vivisection. Berdoe,
    A Catechism of Vivisection, 4.

10. Hughes and Black, "The Preference of Domestic Hens for Different Types of Battery
    Cage Floor."

11. Dawkins, "Cage Size and Flooring Preferences in Litter-Reared and Cage-Reared
    Hens."

12. Zulkifli and Khatijah, "The Relationships Between Cage Floor Preferences and Per-
    formance in Broiler Chickens."

13. P. Townsend (1997), cited in E. G. Patterson-Kane, D. N. Harper, and M. Hunt, "The
    Cage Preferences of Laboratory Rats," *Laboratory Animals* 35 (2001): 74–79.

14. Duncan, "Science-Based Assessment of Animal Welfare."

15. Fraser and Nicol, "Preference and Motivation Research," 185.

16. Ian Duncan, "Animal Welfare: The Question Is 'Do They Suffer?,'" lecture abstract,
    2015, http://www.dierenwelzijn.info/doc/Sprekers_Duncan_kort.pdf.

17. Dawkins, *Why Animals Matter*, 112.

18. Quoted in Denise Cahill, "Change Thinking to Keep Animal Welfare on Agenda,"
    Science Network, http://www.sciencewa.net.au/topics/social-science/item/2397
    -change-thinking-to-keep-animal-welfare-on-agenda.html.

19. Mench and Swanson, "Developing Science-Based Animal Welfare Guidelines."

20. Ibid., 5.

CHAPTER 3   THE ANIMALS WHOM WE EAT

1. Industrial farming is also misleadingly called "factory farming." The use of the term "farm" or "farming" in these expressions (industrial farming, intensive farming, factory farming) euphemizes a practice that is far from bucolic and involves the heinous abuse of animals.

2. Brandt and Aaslyng, "Welfare Measurements of Finishing Pigs on the Day of Slaughter."

3. "The Secret Lives of Chickens," Animals Australia, http://www.makeitpossible .com/features/secret-lives-of-chickens.php.

4. Sy Montgomery, "Are Your Chickens Talking About You?," *Boston Globe*, April 11, 2016, https://www.bostonglobe.com/lifestyle/2016/04/10/are-your-chickens-talking -about-you/xX5DPEGJxAOt5npMAa4FGO/story.html.

5. Maybe the chickens are not really missing out on significant protections. "Humane slaughter" laws don't amount to much. They are routinely violated and underenforced; see Friedrich, "When the Regulators Refuse to Regulate." That "chickens and turkeys make up 98% of slaughtered land animals" comes from Friedrich's earlier essay, "Still in the Jungle."

6. Duncan and Hughes, "Free and Operant Feeding in Domestic Fowls."

7. Bergeron et al., "Stereotypic Oral Behavior in Captive Ungulates."

8. Ibid., 24.

9. Humane Society of the United States, *HSUS Report*.

10. The story of Leo the pig was written up by Bill Crain in "Farm Sanctuary Welcomes Potbellied Pig," *Poughkeepsie Journal*, May 12, 2016, http://www.poughkeepsie journal.com/story/tech/science/environment/2016/05/12/valley-environment-safe -haven-farm-sanctuary/84286628/.

11. Many photos and videos of the animals at Happy Mama Acre are posted on Pinterest: https://www.pinterest.com/joan_hobbs/the-animals-at-happy-mama-acre/. Videos of Doink are available on Google Drive: https://drive.google.com/folderview?id =0B6kJ2N7TnASdZEZYWHBSVVpHMVE&usp=drive_web.

12. Murphy, Nordquist, van der Staay, "A Review of Behavioural Methods to Study Emotion and Mood in Pigs."

13. "The Hidden Lives of Pigs," Compassionate Carnivore, http://www.compassionate -carnivores.org/easy-pig.html.

14. Jennifer Chrisman, "Hail to the Pig: National Pig Day," Fort Gordon Family and MWR, http://www.fortgordon.com/national-pig-day.

15. Neave et al., "Pain and Pessimism."

16. Proctor and Carder, "Can Ear Posture Reliably Measure the Positive Emotional State of Cows?"

17. Sandem, Braastad, and Bøe, "Eye White May Indicate Emotional State on a Frustration-Contentedness Axis in Dairy Cows."

18. Proctor and Carder, "Nasal Temperatures in Dairy Cows Are Influenced by Positive Emotional State."

19. For a review, see Waiblinger et al., "Assessing the Human-Animal Relationship in Farmed Species."

20. Norwood and Lusk, *Compassion, by the Pound*, 98.

21. Ibid.

22. Ibid., 99.

23. Harrison, *Animal Machines*, 37.

24. Quoted in Roberto A. Ferdman, "Why a Top Animal Science Expert Is Worried About the Milk Industry," *Washington Post*, April 21, 2016, http://www.washington post.com/news/wonk/wp/2016/04/21/look-at-what-weve-done-to-cows.

25. *Meat Science*, on the Elsevier website: http://www.journals.elsevier.com/meat -science.

26. Gibson et al., "Pathophysiology of Penetrating Captive Bolt Stunning in Alpacas," abstract.

27. Ferguson and Warner, "Have We Underestimated the Impact of Preslaughter Stress on Meat Quality in Ruminants?," 12.

28. Marks, "A Is for Animal," 13.

29. Nicholas Kristof, "To Kill a Chicken," *New York Times*, March 14, 2015, http:// www.nytimes.com/2015/03/15/opinion/sunday/nicholas-kristof-to-kill-a-chicken .html.

30. Grandin, *Humane Livestock Handling*, 208.

31. Ibid., 207.

32. There might be important trade-offs, however. One of the main environmental problems of meat production is the use of arable land to graze animals, often at the expense of forests and other wild ecosystems. Giving farm animals more space would put increasing pressure on competing uses of the land. The obvious fix, of course, is for humans to dramatically reduce the amount of meat consumed.

33. Millar, "Respect for Animal Autonomy in Bioethical Analysis."

34. Francis Law, "What Do 'Free Range,' 'Organic,' and Other Chicken Labels Really Mean?" *Salon*, January 20, 2011, http://www.salon.com/2011/01/20/what_chicken _labels_really_mean.

35. Humane Society of the United States, *An HSUS Report*, 6.

36. Zonderland et al., "Measuring a Pig's Preference for Suspended Toys."

37. Nicholas Kristof, "Animal Cruelty or the Price of Dinner?," *New York Times*, April 16, 2016, http://www.nytimes.com/2016/04/17/opinion/sunday/animal-cruelty-or-the -price-of-dinner.html.

38. Bertenshaw and Rowlinson, "Exploring Stock Managers' Perceptions of the Human-Animal Relationship."

39. Lürzel et al., "The Influence of Gentle Interactions on Avoidance Distance."

40. Caroline Abels, "Humane Weaning Is Better for Animals and the Bottom Line, Say Farmers," Civil Eats, January 19, 2016, http://civileats.com/2016/01/19/humane -weaning.

41. Scott Timberg, "The Truth About 'Cecil the Lion' Outrage: Why We're So Angry— and What It Says About Us," *Salon*, July 31, 2015, http://www.salon.com/2015/08/01 /the_truth_about_cecil_the_lion_outrage_why_were_so_angry_%E2%80%94 _and_what_it_says_about_us.

42. Bastian et al., "Don't Mind Meat?"

43. European Animal Welfare Platform, http://www.animalwelfareplatform.eu/Welfare -Quality-project.php.

CHAPTER 4   FAT RATS AND LAB CATS

1. Haynes, *Animal Welfare*, xii.
2. Russell and Burch, *The Principles of Humane Experimental Technique*, 12.
3. Ibid., 15.
4. Ibid., 6.
5. Ibid., 3.
6. See Julie Watson, *Animal Welfare* (Baltimore, MD: Johns Hopkins University, Center for Alternatives to Animal Testing, 2010), http://ocw.jhsph.edu/courses/Humane Science/PDFs/CAATLecture8.pdf.
7. Haynes, *Animal Welfare*, 72.
8. Szentágothai, "The 'Brain-Mind' Relation," 323.
9. See, for example, Marc Bekoff, "Do 'Smarter' Dogs Really Suffer More Than 'Dumber' Mice?," *Psychology Today*, April 7, 2013, http://www.psychologytoday.com/blog/animal-emotions/201304/do-smarter-dogs-really-suffer-more-dumber-mice.
10. For a review, see Makowska and Weary, "Assessing the Emotions of Laboratory Rats."
11. See, for example, Jahng et al., "Hyperphagia and Depression-Like Behavior by Adolescence Social Isolation in Female Rats."
12. Jesse Bering, "Rats Laugh, But Not Like Humans," *Scientific American,* July 1, 2012, http://www.scientificamerican.com/article/rats-laugh-but-not-like-human/.
13. Sato et al., "Rats Demonstrate Helping Behavior Toward a Soaked Conspecific."
14. From the *Federal Register*: "We are amending the Animal Welfare Act (AWA) regulations to reflect an amendment to the Act's definition of the term animal. The Farm Security and Rural Investment Act of 2002 amended the definition of animal to specifically exclude birds, rats of the genus *Rattus*, and mice of the genus *Mus*, bred for use in research" (vol. 69, no. 108, June 4, 2004).
15. Würbel, "The Motivational Basis of Caged Rodents' Stereotypies," 88.
16. Mason and Rushen, *Stereotypic Animal Behaviour*, 326.
17. This is as true for humans as it is for animals. Recent attempts to ban or curtail the use of solitary confinement of prisoners are based on the recognition that it can result in severe and irreversible damage to the human brain.
18. Würbel, "The Motivational Basis," 109.
19. Ibid., 110.
20. Reinhardt and Reinhardt, *Variables, Refinement and Environmental Enrichment for Rodents and Rabbits*, 3–4.
21. Ibid., 7.
22. Swaisgood and Shepherdson, "Environmental Enrichment as a Strategy for Mitigating Stereotypies in Zoo Animals," 261–65.
23. Baumans, "Environmental Enrichment," 1251.
24. National Research Council, *Guide for the Care and Use of Laboratory Animals*, 8th ed. (Washington, DC: National Academies Press, 2011), https://grants.nih.gov/grants/olaw/Guide-for-the-Care-and-use-of-laboratory-animals.pdf.
25. Lewis et al., "The Neurobiology of Stereotypy 1," 202.
26. Ibid.

27. Organisation for Economic Co-operation and Development, "Humane Endpoints," 2015, https://www.humane-endpoints.info/en/oecd.
28. See, for example, King, *How Animals Grieve.*
29. Hawkins et al., "Newcastle Consensus Meeting on Carbon Dioxide Euthanasia of Laboratory Animals."
30. Carbone, "Euthanasia and Laboratory Animal Welfare," 161.
31. Quoted in Daniel Cressey, "Imaging Animals for Better Research," *Nature*, June 29, 2011, http://www.nature.com/news/2011/110629/full/news.2011.391.html.
32. Hildebrandt et al., "Anesthesia and Other Considerations for in Vivo Imaging of Small Animals."
33. University of Minnesota, Research Animal Resources, "Restraint and Handling of Animals," 2006, http://www.ahc.umn.edu/rar/handling.html.
34. Markowitz and Eckert, "Giving Power to Animals," 201.
35. Seligman, Maier, and Geer, "Alleviation of Learned Helplessness in the Dog," 256.
36. Coleman et al., "Training Rhesus Macaques for Venipuncture Using Positive Reinforcement Techniques."
37. Markowitz, *Enriching Animal Lives*, 92.
38. Bliss-Moreau, Theil, and Moadab, "Efficient Cooperative Restraint Training with Rhesus Macaques," 99.
39. Reinhardt and Reinhardt, *Variables, Refinement and Environmental Enrichment*, 13.
40. Huang-Brown and Guhad, "Chocolate, an Effective Means of Oral Drug Delivery in Rats."
41. Some experimental protocols with companion dogs still raise questions about stress. For example, one of Horowitz's studies had owners pretend to be angry at their dogs—causing momentary emotional upset. See the bibliography for works by Hare, Horowitz, and Miklosi.
42. Marty Becker's work on Fear Free veterinary care is important.
43. Preilowski, Reger, and Engele, "Combining Scientific Experimentation with Conventional Housing."
44. Beauchamp and Wobber, "Autonomy in Chimpanzees," 117.
45. "Restraint and Handling of Animals," http://www.ahc.umn.edu/rar/handling.html.
46. Savage-Rumbaugh et al., "Welfare of Apes in Captive Environments," 7.
47. Ibid., 9.
48. Ibid., 10.
49. Ibid., 11.
50. Ibid., 11, 12.
51. Ibid., 15.
52. "Jane Goodall," Wikiquote, http://en.wikiquote.org/wiki/Jane_Goodall.
53. The New England Anti-Vivisection Society uses the terms "release" and "restitution" in the context of great ape research. These terms could be interchanged with "refusal" and "rehabilitation." Yet another possibility for the fourth and fifth Rs has been suggested by Shamoo and Resnik in their book *Responsible Conduct of Research*. The fourth R would be *relevance*: "Research protocols that use animals should address questions that have some scientific, medical, or social relevance; all risks/harms to animals need to be balanced against benefits to humans and animals" (226). The fifth R would be *redundancy* (more specifically, its avoidance). "Avoid

redundancy in animal research whenever possible—make sure to do a thorough literature search to ensure that the experiment has not already been done. If it has already been done, provide justification for repeating the work" (226).

54. Institute of Medicine and National Research Council of the National Academies, *Chimpanzees in Biomedical and Behavioral Research: Assessing the Necessity* (Washington, DC: National Academies Press, 2011), http://iom.nationalacademies.org/~/media/Files/Report%20Files/2011/Chimpanzees/chimpanzeereportbrief.pdf; Institute of Medicine, *Council of Councils Working Group on the Use of Chimpanzees in NIH-Supported Research*, 2013, http://dpcpsi.nih.gov/council/pdf/FNL_Report_WG_Chimpanzees.pdf.

55. David Grimm, "Research Chimps to Be Listed as 'Endangered,'" *Science Magazine*, June 12, 2015, http://news.sciencemag.org/people-events/2015/06/research-chimps-be-listed-endangered.

56. Garner, "Stereotypies and Other Abnormal Repetitive Behaviors," 106.

57. Daniel Cressey, "Fat Rats Skew Research Results," *Nature* 464 (2010); Bronwen Martin et al., "'Control' Laboratory Rodents Are Metabolically Morbid: Why It Matters," *Proceedings of the National Academy of Sciences* 107 (2010): 6127–33, doi:10.1073/pnas.0912955107.

58. Baldwin and Bekoff, "Too Stressed to Work," 24. For the original research on "leaky gut" in stressed rats, see Wilson and Baldwin, "Effects of Environmental Stress on the Architecture and Permeability of the Rat Mesenteric Microvasculature," and Wilson and Baldwin, "Environmental Stress Causes Mast Cell Degranulation, Endothelial and Epithelial Changes, and Edema in the Rat Intestinal Mucosa."

59. "Are Lab Mice Too Cold? Why It Matters for Science," *ScienceDaily*, April 19, 2016, http://www.sciencedaily.com/releases/2016/04/160419130014.htm.

60. Institute of Medicine, "Chimpanzees in Biomedical and Behavioral Research: Assessing the Necessity," Report Brief, December 15, 2011, http://iom.nationalacademies.org/Reports/2011/Chimpanzees-in-Biomedical-and-Behavioral-Research-Assessing-the-Necessity/Report-Brief.aspx.

61. Russell and Burch, *The Principles of Humane Experimental Technique*, 158.

62. David Wahlberg, "Controversial UW-Madison Monkey Study Won't Remove Newborns from Mothers," *Wisconsin State Journal*, March 13, 2015, http://host.madison.com/wsj/news/local/education/university/controversial-uw-madison-monkey-study-won-t-remove-newborns-from/article_e8a288f4-5d1a-5ab2-ab50-64b24920b2e3.html.

CHAPTER 5   ZOOED ANIMALS

1. "Killer Whales on Valium: Common Practice?" http://news.discovery.com/animals/whales-dolphins/killer-whales-on-valium-common-practice-140404.htm.

2. Wayne Pacelle, "Breaking News: SeaWorld to End All Orca Breeding," *A Humane Nation* (blog), Humane Society, http://blog.humanesociety.org/wayne/2016/03/seaworld-to-end-orca-breeding.html?credit=blog_post_031716_id8017.

3. "Great Dame: The *Satya* Interview with Gretchen Wyler," June/July 2007, http://www.satyamag.com/jun07/wyler.html.

4. Bradley, *Last Chain on Billie*, 253–54.

5. Association of Zoos and Aquariums, "Accreditation," http://www.aza.org/accredita
    tion/.
6. Detroit Zoological Society Symposium, "From Good Care to Great Welfare."
7. Margodt, *The Welfare Ark*.
8. Brian Handwerk, "National Zoo Deaths: 'Circle of Life' or Animal Care Concerns?"
    National Geographic, December 17, 2013, http://news.nationalgeographic.com/news
    /2013/12/131217-science-zoo-death-smithsonian-gazelle-hog-zebra-kudu-przewalski/.
9. Melfi develops this argument in an important paper: "There Are Big Gaps in Our
    Knowledge, and thus Approach, to Zoo Animal Welfare."
10. Mason, "Using Species Differences in Health and Well-Being to Identify Intrinsic
    Risk and Protective Factors," 2.
11. Bennett, "From Good Care to Great Welfare," 295.
12. Kaufman, "When Babies Don't Fit Plan."
13. Mary Pemberton, "Elephant Not Interested in Using Treadmill," *USA Today*,
    May 16, 2006, http://usatoday30.usatoday.com/tech/science/2006-05-16-elephant
    -treadmill_x.htm. Incidentally, Maggie was eventually moved to the Performing
    Animal Welfare Society Sanctuary in Galt, California, where she is doing well.
14. Föllmi et al., "A Scoring System to Evaluate Physical Condition and Quality of Life
    in Geriatric Zoo Mammals."
15. Hannah Barnes, "How Many Healthy Animals Do Zoos Put Down?" BBC News,
    February 27, 2014, http://www.bbc.com/news/magazine-26356099.
16. See, for example, Hosey, Melfi, and Pankhurst, *Zoo Animals*, 315.
17. Alan Hope, "Antwerp Zoo Introduces No-Kill Policy for Surplus Animals," *Flanders
    Today*, April 13, 2016, http://www.flanderstoday.eu/politics/antwerp-zoo-introduces
    -no-kill-policy-surplus-animals.
18. Mason, "Using Species Differences in Health and Well-Being," 2.
19. "Captive Wildlife Issues: Abnormal Behaviours," Born Free Foundation, http://
    www.bornfree.org.uk/campaigns/zoo-check/captive-wildlife-issues/abnormal-be
    haviours/.
20. Shepherdson et al., "Individual and Environmental Factors Associated with Stereo-
    typic Behavior and Fecal Glucocorticoid Metabolite Levels in Zoo Housed Polar
    Bears."
21. Kurtycz et al., "The Choice to Access Outdoor Areas Affects the Behavior of Great
    Apes."
22. Steve Johnson, "Zoo Innovations Have Animals Foraging," *Chicago Tribune*, March
    20, 2015; Woodland Park Zoo, "Animal Enrichment," http://www.zoo.org/enrich
    ment; Big Cat Rescue, "Vacation Rotation," January 10, 2015, http://bigcatrescue
    .org/vacation-rotation.
23. "Beijing Zoo to Better Protect Animal Rights with New Projects," ChinaDaily.com,
    July 29, 2015, http://www.chinadaily.com.cn/china/2015-06/29/content_21130479
    .htm.
24. Sherwen et al., "Little Penguins."
25. Kelli Grant, "10 Things Zoos Won't Tell You," MarketWatch, May 31, 2011, http://
    www.marketwatch.com/story/guid/1cd1b00c-889a-11e0-9ed9-07212803fad6. On
    civetone and Calvin Klein Obsession, see http://blogs.scientificamerican.com

/thoughtful-animal/youe28099ll-never-guess-how-biologists-lure-jaguars-to-cam era-traps/.

26. Markowitz, *Enriching Animal Lives*, 1.

27. Markowitz and Eckert, "Giving Power to Animals," 203.

28. "What's the Future of Zoos and Aquariums?," *CBCNews*, March 19, 2016, http:// www.cbc.ca/news/technology/zoos-aquariums-conservation-1.3499034.

29. Špinka and Wemelsfelder, "Environmental Challenge and Animal Agency," 28.

30. Ibid., 27.

31. Langbein, Siebert, and Nürnberg, "On the Use of an Automated Learning Device by Group-Housed Dwarf Goats."

32. Špinka and Wemelsfelder, "Environment Challenge and Animal Agency," 4.

33. Ibid., 33.

34. Špinka et al., "Mammalian Play"; Panksepp, "Science of the Brain as a Gateway to Understanding Play."

35. Kagan and Veasey, "Challenges of Zoo Animal Welfare." On giving animals freedom to exert control, see Webster, *Animal Welfare*. On freedom from boredom, see Ryder, *Animal Revolution*.

36. Charles Siebert, "The Dark Side of Zootopia," *New York Times Magazine*, November 18, 2014, http://www.nytimes.com/2014/11/23/magazine/the-dark-side-of-zootopia.html.

37. Taylor, *Respect for Nature*, 173–74.

38. Bekoff, "Naturalizing and Individualizing Animal Well-Being and Animal Minds," 80.

39. Rees, *An Introduction to Zoo Biology and Management*, 330.

40. Jon Cohen, "Zoo Futures," *Conservation Magazine*, March 8, 2013, http://conservation magazine.org/2013/03/zoo-futures/.

41. Kellert, *Kinship to Mastery*, 100. See also Grazian, *American Zoo*.

42. Waldau, *Animal Studies*, 270.

43. Natalia Lima, "Could We Offer Wildlife Encounters Without Holding Animals Captive?," *Care2*, April 16, 2016, http://www.care2.com/causes/could-we-offer-wild life-encounters-without-holding-animals-captive.html.

44. Jamieson, "Against Zoos."

45. Andrew Moss, Eric Jensen, and Marcus Gusset, *A Global Evaluation of Biodiversity Literacy in Zoo and Aquarium Visitors* (Gland, Switzerland: World Association of Zoos and Aquariums, 2014), http://www.waza.org/files/webcontent/1.public_site/5. conservation/un_decade_biodiversity/WAZA%20Visitor%20Survey%20Report.pdf.

46. Marc Bekoff, "The Conservation Charade: US Zoos Propose Importing Wild African Elephants," *Huffington Post*, November 9, 2015, http://www.huffingtonpost .com/marc-bekoff/the-conservation-charade-_b_8498884.html.

47. Marc Bekoff, "Secret Flight of Swaziland Elephants Avoids Legal Challenge," *Psychology Today*, March 13, 2016, http://www.psychologytoday.com/blog/animal -emotions/201603/secret-flight-swaziland-elephants-avoids-legal-challenge.

48. Kagan, Allard, and Carter, "Exotic Animal Welfare," 26.

49. "Phasing Out Elephants?," Elephants in Canada website, http://www.elephantsin canada.com/other-zoos-closing-elephant-exhibits.

50. Detroit Zoo, http://www.detroitzoo.org/About/zoo-history.

51. Ron Kagan, director of the Detroit Zoo, personal e-mail communication to Marc, February 28, 2015.

52. "Danish Zoo That Killed Marius the Giraffe Puts Down Four Lions," Agence France-Presse, March 25, 2014, http://www.theguardian.com/world/2014/mar/25 /danish-copenhagen-zoo-kills-four-lions-marius-giraffe.

CHAPTER 6   CAPTIVE AND COMPANION

1. See Pierce, *The Last Walk*.

2. See Pierce, *Run, Spot, Run*.

3. On dogs, see Horowitz, *Inside of a Dog*; Miklosi, *Dog Behavior, Evolution, and Cognition*; Hare and Woods, *The Genius of Dogs*; and Rooney and Bradshaw, "Canine Welfare Science." On cats, see Bradshaw, *Cat Sense*, and Bradshaw, "Sociality in Cats."

4. Rooney and Bradshaw, "Canine Welfare Science," 241–74.

5. American Pet Products Association, "Pet Market Size & Ownership Statistics," http://www.americanpetproducts.org/press_industrytrends.asp.

6. See, for example, Olmert, *Made for Each Other*.

7. Purbita Saha, "Do Birds Have an Inherent Right to Fly?," *Audubon* (March/April 2016), http://www.audubon.org/magazine/march-april-2016/do-birds-have-inherent -right-fly.

8. Burghardt, "Environmental Enrichment and Cognitive Complexity in Reptiles and Amphibians."

9. Warwick, "The Morality of the Reptile 'Pet' Trade," 79.

10. Horowitz, "*Canis familiaris*," 16.

11. *Secret Science of the Dog Park* (Toronto, ON: Stornoway Productions, 2015). Documentary film.

12. American Veterinary Society of Animal Behavior, "Position Paper on the Use of Dominance Theory in Behavior Modification of Animals," 2008.

13. Marc Bekoff, "Social Dominance Is Not a Myth: Wolves, Dogs, and Other Animals," *Psychology Today*, February 15, 2012, https://www.psychologytoday.com/blog /animal-emotions/201202/social-dominance-is-not-myth-wolves-dogs-and.

14. Yin, *How to Behave So Your Dog Behaves*.

15. Palmer and Sandøe, "For Their Own Good."

16. Cornwall, "There Will Be Blood."

17. Turner and Bateson, *The Domestic Cat*, 188–90.

18. Palmer and Sandøe, "For Their Own Good," 146, quoting the philosopher Sandra Lee Bartky (1990).

19. Snowdon et al., "Cats Prefer Species-Appropriate Music."

20. McGowan et al., "Positive Affect and Learning."

21. Dognition, http://www.dognition.com.

22. Normando and Gelli, "Behavioral Complaints and Owners' Satisfaction in Rabbits, Mustelids, and Rodents Kept as Pets."

23. Wemelsfelder, "Animal Boredom," 87.

24. Turner and Bateson, *The Domestic Cat*, 195.

25. See, for example, Balcombe, "Cognitive Evidence of Fish Sentience," and Brown, Laland, and Krause, *Fish Cognition and Behaviour*.

26. Salvanes et al., "Environmental Enrichment Promotes Neural Plasticity and Cognitive Ability in Fish," 1767.

27. Wemelsfelder et al., "Assessing the 'Whole Animal.'"

28. American College of Animal Welfare, "Animal Welfare Principles," http://www.acaw .org/animal_welfare_principles.html, 2010.

CHAPTER 7   BORN TO BE WILD?

1. Alexandra Mester, "Three Bears Orphaned in Wild Find Toledo Zoo Home Is Just Right," *Toledo Blade*, March 2, 2106, http://www.toledoblade.com/local/2016/03/02/3 -bears-orphaned-in-wild-find-Toledo-Zoo-home-is-just-right.html.

2. A Google search on catch-and-release mortality rates for fish yields a bounty of information on this topic.

3. Bateson and Bradshaw, "Physiological Effects of Hunting Red Deer."

4. Bryan et al., "Heavily Hunted Wolves Have Higher Stress and Reproductive Steroids Than Wolves with Lower Hunting Pressure."

5. Ordiz et al., "Do Bears Know They Are Being Hunted?," 21.

6. Ibid.

7. Animal and Plant Health Inspection Service, "'Table G: Animals Taken by Wildlife Services—FY 2014," https://www.aphis.usda.gov/wildlife_damage/prog_data /2014/G/Tables/Table%20G_ShortReport.pdf.

8. Alex Burness, "Killing No. 317: The Story of a Boulder Bear Out of Strikes" *Boulder Daily Camera*, September 28, 2015, http://www.dailycamera.com/news/boulder /ci_28881363/killing-no-317-story-boulder-bear-out-strikes?so.

9. Larry Pynn, "BC Conservation Officers Criticized for 'Cavalier' Killing of Predators," *Vancouver Sun*, July 11, 2015, http://www.vancouversun.com/technology/con servation+officers+criticized+cavalier+killing+predators/11209072/story.html.

10. AVMA, "Welfare Implications of Leghold Trap Use in Conservation and Research: Literature Review," April 30, 2008, https://www.avma.org/KB/Resources/Litera tureReviews/Pages/Welfare-Implications-of-Leghold-Trap-Use-in-Conservation -and-Research.aspx.

11. Hervieux et al., "Managing Wolves."

12. Proulx et al., "Poisoning Wolves with Strychnine Is Unacceptable in Experimental Studies and Conservation Programmes."

13. Hervieux et al., "Managing Wolves," abstract.

14. Pearce, *The New Wild*. See also van Dooren, "Invasive Species in Penguin Worlds"; Nagy and Johnson, *Trash Animals*; Ramp and Bekoff, "Compassion as a Practical and Evolved Ethic for Conservation."

15. Researchers distinguish between translocation, reintroduction, restocking, relocation, and repatriation. For simplicity's sake, we use "translocation" and "reintroduction" as interchangeable. Translocations are not necessarily reintroductions, but both involve moving animals around. For a detailed discussion of terminology, see Teixeira et al., "Revisiting Translocation and Reintroduction Programmes."

16. Germano et al., "Mitigation-Driven Translocations."

17. Devan-Song, "Ecology and Conservation of the Bamboo Pit Viper."

18. Cornwall, "There Will Be Blood."

19. Shelby Sebens, "US Culls Over 1,200 Oregon Cormorants, Sparks Outcry," Reuters, September 25, 2015, http://www.reuters.com/article/us-usa-cormorants-idUSKC N0RP2CQ20150925.

20. Jeff Thompson, "Portland Audubon Calls for End to Cormorant-Killing Program After Colony Collapse," KGW.com, May 20, 2016, http://www.kgw.com/news/local /animal/audubon-columbia-river-cormorant-colony-has-collapsed/207237506.

21. Marc Bekoff, "Killing Barred Owls to Save Spotted Owls? Problems from Hell," *Psychology Today*, November 1, 2014, http://www.psychologytoday.com/blog/animal -emotions/201411/killing-barred-owls-save-spotted-owls-problems-hell.

22. Cornwall, "There Will Be Blood."

23. William Lynn, "Barred Owls in the Pacific Northwest: An Ethics Brief," 2012, http:// www.williamlynn.net/pdf/lynn-2011-barred-owls.pdf.

24. "Rewilding," Save China's Tigers, http://savechinastigers.org/rewilding.html.

25. Bekoff, "Field Studies and Animal Models."

26. Mallonée and Joslin, "Traumatic Stress Disorder Observed in an Adult Wild Captive Wolf."

27. Minteer et al., "Avoiding (Re)extinction."

28. Evenhuis and Marshall, "New Species Without Dead Bodies."

29. Rachel E. Gross, "Ridiculously Gorgeous Rare Bird Finally Photographed," *Slate,* September 24, 2015, http://www.slate.com/blogs/wild_things/2015/09/24/mustached _kingfisher_photographed_for_first_time_proves_it_is_definitely.html.

30. Chris Filardi, "Finding Ghosts," *Field Journal* (blog), American Museum of Natural History, September 23, 2015, http://www.amnh.org/explore/news-blogs /from-the-field-posts/field-journal-finding-ghosts?utm_source=social-media&utm _medium=twitter&utm_term=20150923-wed&utm_campaign=expedition.

31. Ernest et al., "Fractured Genetic Connectivity Threatens a Southern California Puma"; Vickers et al., "Survival and Mortality of Pumas (*Puma concolor*).

32. Bunkley et al., "Anthropogenic Noise Alters Bat Activity Levels and Echolocation Calls."

33. Owen et al., "Hearing Sensitivity in Context."

34. Barber et al., "The Costs of Chronic Noise Exposure for Terrestrial Organisms."

35. Sandrine Ceurstemont, "Ocean Commotion: Protecting Sea Life from Our Noise," *New Scientist,* April 8, 2015, https://www.newscientist.com/article/mg22630161-200 -ocean-commotion-protecting-sea-life-from-our-noise/.

36. McDonald, Hildebrand, and Wiggins, "Increases in Deep Ocean Ambient Noise."

37. Dyndo et al., "Harbour Porpoises React to Low Levels of High Frequency Vessel Noise."

38. Simpson et al., "Anthropogenic Noise Increases Fish Mortality by Predation."

39. Elisabeth Malkin and Paulina Villegas, "Tourists Thwart Turtles from Nesting in Costa Rica," *New York Times,* September 18, 2015, http://www.nytimes.com/2015/09/19 /world/americas/tourists-thwart-turtles-from-nesting-in-costa-rica.html.

40. Penny Sarchet, "Antarctic Tourism May Pose Disease Threat to Penguins," *New Scientist,* December 19, 2014, http://www.newscientist.com/article/dn26725-antarctic -tourism-may-pose-disease-threat-to-penguins.html.

41. Grimaldi et al., "Infectious Diseases of Antarctic Penguins."

42. Ferris Jabr, "Animals Spy a New Enemy: Drones," *New York Times*, October 20,

2015. Original research conducted by Ditmer et al., "Bears Show a Physiological but Limited Behavioral Response to Unmanned Aerial Vehicles."

43. Bean et al., "Behavioural and Physiological Responses of Birds to Environmentally Relevant Concentrations of an Antidepressant."

44. Brodin et al., "Ecological Effects of Pharmaceuticals in Aquatic Systems."

45. Hedgespeth et al., "Ecological Implications of Altered Fish Foraging After Exposure to an Antidepressant Pharmaceutical."

46. Philip Hoare, "Whales Are Starving—Their Stomachs Full of Our Plastic Waste," *Guardian*, March 30, 2016, http://www.theguardian.com/commentisfree/2016/mar /30/plastic-debris-killing-sperm-whales.

47. For a refreshingly novel approach to conservation that recognizes the importance of individuals in wildlife rehabilitation, see Aitken, *A New Approach to Conservation.*

48. Kartick Satyanarayan and Geeta Seshamani, "Elephants, Bears and Monkeys in India: Welfare and Conservation Both Served," http://www.bornfree.org.uk/file admin/user_upload/files/compcon/Draft_Abstract_booklet.pdf, 25. Many other examples can be found in the abstracts included in this document.

49. Raman, "Leopard Landscapes."

50. Institute for Humane Education, "The World Becomes What You Teach," TEDtalk by Zoe Weil, http://humaneeducation.org/blog/2012/06/05/the-world-becomes-what -you-teach-2.

51. For an interesting discussion of wildlife in the Anthropocene, see Lorimer, *Wildlife in the Anthropocene.*

52. Deryabina et al., "Long-Term Census Data Reveal Abundant Wildlife Populations at Chernobyl."

53. Lisa Brady, "How Wildlife Is Thriving in the Korean Peninsula's Demilitarized Zone," *Guardian*, April 13, 2012, http://www.theguardian.com/environment/2012/apr /13/wildlife-thriving-korean-demilitarised-zone; Annie Snider, "Outside Guantanamo's Prisons, 'It's Really a Biologist's Dream,'" *New York Times*, June 17, 2011, http:// www.nytimes.com/gwire/2011/06/17/17greenwire-outside-guantanamos-prisons-its -really-a-biolo-72463.html?pagewanted=all.

CHAPTER 8   COEXISTENCE IN THE ANTHROPOCENE AND BEYOND

1. Kathleen McLaughlin, "China Finally Setting Guidelines for Treating Lab Animals," *Science*, March 21, 2016, http://www.sciencemag.org/news/2016/03/china-finally -setting-guidelines-treating-lab-animals; Richard Denison, "Historic Deal on TSCA Reform Reached, Setting Stage for a New Law After 40 Years of Waiting," *EDF Health*, May 23, 2016, http://blogs.edf.org/health/2016/05/23/historic-deal-on-tsca -reform-reached-setting-stage-for-a-new-law-after-40-years-of-waiting/; Editorial Board, "Ban Animal Use in Military Medical Training," *New York Times*, June 25, 2016, http://www.nytimes.com/2016/06/26/opinion/ban-animal-use-in -military-medical-training.html?emc=edit_tnt_20160625&nlid=67348256&tnte mailo=y&_r=0; Uke Goñi, "Buenos Aires Zoo to Close After 140 Years: 'Captivity Is Degrading,'" *Guardian*, June 23, 2016, https://www.theguardian.com/world/2016 /jun/23/buenos-aires-zoo-close-animals-captivity-argentina?CMP=twt_a-world_b -gdnworld; Amanda Lindner, "Oh Yeah! Iran Bans Use of Wild Animals in Circuses!," One Green Planet, March 30, 2016, http://www.onegreenplanet.org/news

/iran-bans-use-of-wild-animals-in-circuses; Humane Society International, "More than 42 Airlines Adopt Wildlife Trophy Bans after Cecil the Lion's Death," August 27, 2015, http://www.hsi.org/news/press_releases/2015/08/42-airlines-adopt-wildlife -trophy-bans-082715.html.

2. Broom and Fraser, *Domestic Animal Behaviour and Welfare*, p. 6, our italics.

3. Ibid., 14.

4. Mellor, "Updating Animal Welfare Thinking."

5. Yuval Noah Harari, "Industrial Farming Is One of the Worst Crimes in History," *Guardian*, September 25, 2015.

6. Broom and Fraser, *Domestic Animal Behaviour and Welfare*, 6.

7. Bekoff, *Rewilding Our Hearts*.

8. Katherine Martinko, "Children Spend Less Time Outside Than Prison Inmates," *TreeHugger*, March 25, 2016, http://www.treehugger.com/culture/children-spend -less-time-outside-prison-inmates.html.

9. Frans de Waal, "Who Apes Whom?" *New York Times,* September 15, 2015.

10. *Atlantic,* June 2015, http://www.theatlantic.com/magazine/archive/2015/06/the-big -question/392108.

11. The theory of the adjacent possible was first proposed by biophysicist Stuart Kauffman in 2002, but Johnson is the first to apply the concept to creative thought.

12. Steven Johnson, "The Genius of the Tinkerer," *Wall Street Journal*, September 25, 2010, http://www.wsj.com/articles/SB10001424052748703989304575503730101860 838. See also Johnson, *Where Good Ideas Come From*.

Bibliography

Agrawal, H. C., M. W. Fox, and W. A. Himwich. "Neurochemical and Behavioral Effects of Isolation-Rearing in the Dog." *Life Sciences* 6, no. 1 (1967): 71–78.

Aitken, Gill. *A New Approach to Conservation: The Importance of the Individual Through Wildlife Rehabilitation.* Burlington, VT: Ashgate, 2004.

Allen, Colin. "Fish Cognition and Consciousness." *Journal of Agricultural and Environmental Ethics* 26 (2013): 25–39.

Amat, Marta, Tomàs Camps, and Xavier Manteca. "Stress in Owned Cats: Behavioural Changes and Welfare Implications." *Journal of Feline Medicine and Surgery* (June 22, 2015): 1–10. doi:10.1177/1098612X15590867.

Amiot, C., and B. Bastain. "Toward a Psychology of Human-Animal Relations." *Psychological Bulletin* (2014): 1–42.

Andersen, Kip, and Keegan Kuhn. *The Sustainability Secret: Rethinking Our Diet to Transform the World.* San Rafael, CA: Earth Aware, 2015.

Appleby, Michael C. "Do Hens Suffer in Battery Cages? A Review of the Scientific Evidence Commissioned by the Athene Trust." Institute of Ecology and Resource Management, University of Edinburgh. October 1991.

Appleby, Michael, and Barry Hughes, eds. *Animal Welfare.* 2nd ed. Wallingford, UK: CABI, 2011.

Aydinonat, Denise, et al. "Social Isolation Shortens Telomeres in African Grey Parrots (*Psittacus erithacus erithacus*)." *PLOS One* (April 4, 2014). doi:10.1371/journal.pone.0093839.

Balcombe, Jonathan. "Cognitive Evidence of Fish Sentience." *Animal Sentience* 1, no. 2 (2016).

———. *What a Fish Knows: The Inner Lives of Our Underwater Cousins.* New York: Scientific American/Farrar, Straus and Giroux, 2016.

Baldwin, Ann, and Marc Bekoff. "Too Stressed to Work." *New Scientist* 9 (2007): 24, http://www.ip.usp.br/portal/images/stories/cepa/Too%20stressed%20work.pdf.

Barber, Jesse R., Kevin R. Crooks, and Kurt M. Fristrup. "The Costs of Chronic Noise Exposure for Terrestrial Organisms." *Trends in Ecology & Evolution* 25, no. 3 (2010): 180–89.

Barnett, J. L., et al. "A Review of the Welfare Issues for Sows and Piglets in Relation to Housing." *Australian Journal of Agricultural Research* 52 (2001): 1–28.

Baruch-Mordo, Sharon, et al. "Stochasticity in Natural Forage Production Affects Use of Urban Areas by Black Bears: Implications to Management of Human-Bear Conflicts." *PLOS ONE* 9 (January 8, 2014): e85122. doi:10.1371/journal.pone.0085122.

Bastian, Brock, et al. "Don't Mind Meat? The Denial of Mind to Animals Used for Human Consumption." *Personality and Social Psychology Bulletin* 38, no. 2 (2012): 247–56.

Bateson, Patrick, and Elizabeth L. Bradshaw. "Physiological Effects of Hunting Red Deer (*Cervus elaphus*)." *Proceedings of the Royal Society B.* 264 (1997): 1707–14.

Baumans, V. "Environmental Enrichment: A Right of Rodents!" In *Progress in the Reduction, Refinement, and Replacement of Animal Experimentation: Proceedings of the Third World Congress on Alternatives and Animal Use,* edited by M. Balls, A.-M. van Zeller, and M. Halder, 1251–55. Amsterdam: Elsevier, 2000.

Baxter, M. "The Welfare Problems of Laying Hens in Battery Cages." *Veterinary Record* 134 (1994): 614–19.

Bean, Tom G., et al. "Behavioural and Physiological Responses of Birds to Environmentally Relevant Concentrations of an Antidepressant." *Philosophical Transactions of the Royal Society B: Biological Sciences* 369 (2014). doi:10.1098/rstb.2013.0575.

Beauchamp, Tom L., and Victoria Wobber. "Autonomy in Chimpanzees." *Theoretical Medicine and Bioethics* 35 (2014): 117–32.

Beck, Benjamin. *Thirteen Gold Monkeys.* Parker, CO: Outskirts Press, 2013.

Bekoff, Marc. *The Animal Manifesto: Six Reasons for Expanding Our Compassion Footprint.* Novato, CA: New World Library, 2010.

———. *The Emotional Lives of Animals: A Leading Scientist Explores Animal Joy, Sorrow, and Empathy—and Why They Matter.* Novato, CA: New World Library, 2007.

———. "Field Studies and Animal Models: The Possibility of Misleading Inferences." In *Progress in the Reduction, Refinement and Replacement of Animal Experimentation,* edited by M. Balls, A.-M. van Zeller, and M. E. Halder, 1553–59. Amsterdam: Elsevier, 2000.

———, ed. *Ignoring Nature No More: The Case For Compassionate Conservation.* Chicago: University of Chicago Press, 2013.

———. *Minding Animals: Awareness, Emotions, and Heart.* New York: Oxford University Press, 2002.

———. "Naturalizing and Individualizing Animal Well-Being and Animal Minds: An Ethologist's Naivete Exposed?" In *Wildlife Conservation, Zoos, and Animal Protection: A Strategic Analysis,* edited by Andrew N. Rowan, 63–129. North Grafton, MA: Tufts Center for Animals and Public Policy, 1995.

———. *Rewilding Our Hearts: Building Pathways of Compassion and Coexistence.* Novato, CA: New World Library, 2014.

———. *Why Dogs Hump and Bees Get Depressed: The Fascinating Science of Animal Intelligence, Emotions, Friendship, and Conservation.* Novato, CA: New World Library, 2013.

Bekoff, Marc, and Jessica Pierce. *Wild Justice: The Moral Lives of Animals.* Chicago: University of Chicago Press, 2009.

Bennett, Cynthia. "From Good Care to Great Welfare." *Journal of Applied Animal Welfare Science* 16 (2013): 295–99.

Berdoe, Edward. *A Catechism of Vivisection: The Whole Controversy Argued in All Its Detail*. London: Swan Sonnenschein, 1903.

Berger, Joel, et al. "Legacies of Past Exploitation and Climate Affect Mammalian Sexes Differently on the Roof of the World—The Case of Wild Yaks." *Scientific Reports* 5 (2015). doi: 10.1038/srep08676.

Bergeron, R., et al. "Stereotypic Oral Behavior in Captive Ungulates: Foraging, Diet and Gastrointestinal Function." In Mason and Rushen, *Stereotypic Animal Behaviour*, 19–25.

Berns, Gregory S., Andrew M. Brooks, and Mark Spivak. "Functional MRI in Awake Unrestrained Dogs." *PLOS One* 7 (2012): e38027. doi:10.1371/journal.pone.0038027.

Bertenshaw, Catherine, and Peter Rowlinson. "Exploring Stock Managers' Perceptions of the Human-Animal Relationship on Dairy Farms and an Association with Milk Production." *Anthrozoos* 22, no. 1 (2009): 59–69.

Bessei, W. "Welfare of Broilers: A Review." *World's Poultry Science Journal* 62 (2006): 455–66.

Bliss-Moreau, Eliza, Jacob H. Theil, and Gilda Moadab. "Efficient Cooperative Restraint Training with Rhesus Macaques." *Journal of Applied Animal Welfare Science* 16 (2013): 98–117.

Boissy, A., et al. "Cognitive Sciences to Relate Ear Postures to Emotions in Sheep." *Animal Welfare* 20 (2011): 47–56.

Boyle, L. A., et al. "Effect of Gestation Housing on Behaviour and Skin Lesions of Sows in Farrowing Crates." *Applied Animal Behaviour Science* 76 (2002): 119–34.

Bradley, Carol. *Last Chain on Billie: How One Extraordinary Elephant Escaped the Big Top*. New York: St. Martin's, 2014.

Bradshaw, A. L., and A. Poling. "Choice by Rats for Enriched Versus Standard Home Cages: Plastic Pipes, Wood Platforms, Wood Chips, and Paper Towels as Enrichment Items." *Journal of the Experimental Analysis of Behaviour* 55 (1991): 245–50.

Bradshaw, John W. S. *Cat Sense: How the New Feline Science Can Make You a Better Friend to Your Pet*. New York: Basic Books, 2014.

———. "Sociality in Cats: A Comparative Review." *Journal of Veterinary Behavior* 11 (January–February 2015): 113–24.

Brambell, Roger. *Report of the Technical Committee to Enquire into the Welfare of Animals Kept Under Intensive Livestock Husbandry Systems*. London: Her Majesty's Stationary Office, 1965.

Brandt, P., and M. D. Aaslyng. "Welfare Measurements of Finishing Pigs on the Day of Slaughter: A Review." *Meat Science* 103 (2015): 13–23.

Braverman, Irus. *Wild Life: The Institution of Nature*. Palo Alto, CA: Stanford University Press, 2015.

———. *Zooland: The Institution of Captivity*. Palo Alto, CA: Stanford Law Books, 2012.

Briefer, Elodie F., Federico Tettamanti, and Alan G. McElligott. "Emotions in Goats: Mapping Physiological, Behavioural and Vocal Profiles." *Animal Behaviour* 99 (2015): 131–43.

Brodin, T., et al. "Dilute Concentrations of a Psychiatric Drug Alter Behavior of Fish from Natural Populations." *Science* 339 (2013): 814–15.

————. "Ecological Effects of Pharmaceuticals in Aquatic Systems—Impacts Through Behavioural Alterations." *Philosophical Transactions of the Royal Society B.* 369 (2014). doi:10.1098/rstb.2013.0580.

Broom, Donald. "Cognitive Ability and Awareness in Domestic Animals and Decisions About Obligations to Animals." *Applied Animal Behaviour Science* 126 (2010): 1–11.

Broom, Donald M., and Andrew F. Fraser. *Domestic Animal Behaviour and Welfare.* 4th ed. Cambridge, MA: CABI, 2007.

Brown, Culum, Kevin Laland, and Jens Krause, eds. *Fish Cognition and Behaviour.* 2nd ed. Hoboken, NJ: Wiley-Blackwell, 2011.

Bryan, Heather M., et al. "Heavily Hunted Wolves Have Higher Stress and Reproductive Steroids Than Wolves with Lower Hunting Pressure." *Functional Ecology,* November 12, 2014. doi:10.1111/1365-2435.12354.

Bunkley, Jessie P., et al. "Anthropogenic Noise Alters Bat Activity Levels and Echolocation Calls." *Global Ecology and Conservation* 3 (2015): 62–71.

Burghardt, Gordon. "Environmental Enrichment and Cognitive Complexity in Reptiles and Amphibians: Concepts, Review, and Implications for Captive Populations." *Applied Animal Behaviour Science* 147 (2013): 286–98.

Bush, Emma, Sandra E. Baker, and David W. MacDonald. "Global Trade in Exotic Pets 2006–2012." *Conservation Biology* 28 (2014): 663–76.

Capucille, D. J., M. H. Poore, and G. M. Rogers. "Castration in Cattle: Techniques and Animal Welfare Issues." *Compendium* 24 (2002): 66–71.

Carbone, Larry. "Euthanasia and Laboratory Animal Welfare." In *Laboratory Animal Welfare,* edited by Kathryn Bayne and Patricia V. Turner, 157–69. Waltham, MA: Academic Press, 2014.

Carneiro, Manuela, et al. "Assessment of the Exposure to Heavy Metals in Griffon Vultures (*Gyps fulvus*) from the Iberian Peninsula." *Ecotoxicology and Environmental Safety* 113 (2015): 295–301. doi:10.1016/j.ecoenv.2014.12.016.

Christiansen, Stine Billeschou, and Bjorn Forkman. "Assessment of Animal Welfare in a Veterinary Context—A Call for Ethologists." *Applied Animal Behaviour Science* 106 (2007): 203–20.

Clark, Jonathan. "Labourers or Lab Tools? Rethinking the Role of Lab Animals in Clinical Trials." In *The Rise of Critical Animal Studies: From the Margins to the Centre,* edited by Nik Taylor and Richard Twine, 139–64. New York: Routledge, 2014.

Clive, J. C. Phillips. *The Animal Trade.* Cambridge, MA: CABI, 2015.

Coleman, Kristine, et al. "Training Rhesus Macaques for Venipuncture Using Positive Reinforcement Techniques: A Comparison with Chimpanzees." *Journal of the American Association for Laboratory Animal Science* 47 (2008): 37–41.

Cooper, Jonathan J., et al. "The Welfare Consequences and Efficacy of Training Pet Dogs with Remote Electronic Training Collars in Comparison to Reward Based Training." *PLOS One* 9 (2014): e102722. doi: 10.1371/journal.pone.0102722.

Cornwall, Warren. "There Will Be Blood," *Conservation Magazine,* October 24, 2014. http://conservationmagazine.org/2014/10/killing-for-conservation.

Cressey, Daniel. "Fat Rats Skew Research Results." *Nature* 464 (2010).

Dawkins, M. S., and Hardie, S. "Space Needs of Laying Hens." *British Poultry Science* 30 (1989): 413–16.

Dawkins, Marian S. "Cage Height Preference and Use in Battery-Kept Hens." *Veterinary Record* 116 (1985): 345–47.

———. "Cage Size and Flooring Preferences in Litter-Reared and Cage-Reared Hens." *British Poultry Science* 24 (1983): 177–82.

———. "Using Behaviour to Assess Animal Welfare." *Animal Welfare* (2004): S3–S7.

———. *Why Animals Matter: Animal Consciousness, Animal Welfare, and Human Well-Being.* New York: Oxford University Press, 2012.

Deryabina, T. G., et al. "Long-Term Census Data Reveal Abundant Wildlife Populations at Chernobyl." *Current Biology* 25 (2015): R811–R826.

Detroit Zoological Society Symposium. "From Good Care to Great Welfare: Selected Papers from the Detroit Zoological Society Symposium (August 2011)." *Journal of Applied Animal Welfare Science* 16, no. 4 (2013).

Devan-Song, Elizabeth Anne. "Ecology and Conservation of the Bamboo Pit Viper: Natural History, Demography and the Effects of Translocation." Master's thesis, University of Rhode Island, 2014. http://digitalcommons.uri.edu/theses/456.

Ditmer, Mark A., et al. "Bears Show a Physiological but Limited Behavioral Response to Unmanned Aerial Vehicles." *Current Biology* 25 (2015): 2278–83.

D'Silva, Joyce. "Adverse Impact of Industrial Animal Agriculture on the Health and Welfare of Farmed Animals." *Integrative Zoology* 1 (2006): 53–58.

Duncan, I. J. H. "Animal Welfare Issues in the Poultry Industry: Is There a Lesson to Be Learned?" *Journal of Applied Animal Welfare Science* 4 (2001): 207–21.

———. "Science-Based Assessment of Animal Welfare: Farm Animals." *Revue Scientifique et Technique-Office International Des Epizooties* 24 (2005): 483–92.

Duncan, Ian, and Barry Hughes. "Free and Operant Feeding in Domestic Fowls." *Animal Behaviour* 20 (1972): 775–77.

Dyndo, Monika, et al. "Harbour Porpoises React to Low Levels of High Frequency Vessel Noise." *Scientific Reports* 5 (2015).

Eadie, Edward N. *Understanding Animal Welfare: An Integrated Approach.* New York: Springer, 2012.

Edgar, J. L., et al. "Measuring Empathic Responses in Animals." *Applied Animal Behaviour Science* 138 (2012): 182–93.

Ernest, Holly, et al. "Fractured Genetic Connectivity Threatens a Southern California Puma (*Puma concolor*) Population." *PLOS One* 9 (2014): e107985.

Ernst, Katrin, et al. "A Complex Automatic Feeding System for Pigs Aimed to Induce Successful Behavioural Coping by Cognitive Adaptation." *Applied Animal Behaviour Science* 91 (2005): 205–18.

Evenhuis, Neal L., and Stephen A. Marshall. "New Species Without Dead Bodies: A Case for Photo-Based Descriptions, Illustrated by a Striking New Species of *Marleyimyia* Hesse (Diptera, Bombyliidae) from South Africa." *ZooKeys* 525 (2015): 117.

Fenton, Andrew. "On the Need to Redress an Inadequacy in Animal Welfare Science: Toward an Internally Coherent Framework." *Biology and Philosophy* 27 (2012): 73–93.

Ferguson, Drewe M., and Robyn Dorothy Warner. "Have We Underestimated the Impact of Preslaughter Stress on Meat Quality in Ruminants?" *Meat Science* 80 (2008): 12–19.

Flower, F. C., and D. M. Weary. "The Effects of Early Separation on the Dairy Cow and Calf." *Animal Welfare* 12 (2003): 339–48.

———. "Gait Assessment in Dairy Cattle." *Animal* 3 (2009): 87–95.

Föllmi, J., et al. "A Scoring System to Evaluate Physical Condition and Quality of Life in Geriatric Zoo Mammals." *Animal Welfare* 16 (2007).

Francis, Richard C. *Domesticated: Evolution in a Man-Made World.* New York: Norton, 2015.

Fraser, Andrew F. *Feline Behaviour and Welfare.* Wallingford, UK: CABI, 2012.

Fraser, David. *Understanding Animal Welfare: The Science in Its Cultural Context.* Ames, IA: Wiley-Blackwell, 2008.

Fraser, David, and Christine J. Nicol. "Preference and Motivation Research." In Appleby and Hughes, *Animal Welfare,* 183–99.

French, Thomas. *Zoo Story: Life in the Garden of Captives.* New York: Hachette, 2010.

Friedrich, Bruce. "Still in the Jungle: Poultry Slaughter and the USDA." *New York University Environmental Law Journal* 23 (2015): 248–98.

———. "When the Regulators Refuse to Regulate: Pervasive USDA Underenforcement of the Humane Slaughter Act." *Georgetown Law Journal* 104 (2015): 197–227.

Garner, Joseph P. "Stereotypies and Other Abnormal Repetitive Behaviors: Potential Impact on Validity, Reliability, and Replicability of Scientific Outcomes." *ILAR Journal* 46 (2005): 106–17.

Garner, Robert. *A Theory of Justice for Animals: Animal Rights in a Nonideal World.* New York: Oxford University Press, 2013.

Germano, Jennifer M., et al. "Mitigation-Driven Translocations: Are We Moving Wildlife in the Right Direction?" *Frontiers in Ecology and the Environment.* January 16, 2015. doi:10.1890/140137.

Gibson, Troy J., et al. "Pathophysiology of Penetrating Captive Bolt Stunning in Alpacas (*Vicugna pacos*)." *Meat Science* 100 (2015): 227–31.

Gluck, John P., Tony DiPasquale, and F. Barbara Orleans, eds. *Applied Ethics in Animal Research: Philosophy, Regulation, and Laboratory Applications.* West Lafayette, IN: Purdue University Press, 2002.

Goodwin, D., J. W. Bradshaw, and S. M. Wickens. "Paedomorphosis Affects Agonistic Visual Signals of Domestic Dogs." *Animal Behaviour* 53 (1997): 297–304.

Grandin, Temple, with Mark Deesing. *Humane Livestock Handling.* North Adams, MA: Storey, 2008.

Grandin, Temple. "Auditing Animal Welfare at Slaughter Plants." *Meat Science* 86 (2010): 56–65.

———. "The Feasibility of Using Vocalization Scoring as an Indicator of Poor Welfare During Cattle Slaughter." *Applied Animal Behaviour Science* 56 (1998): 121–28.

———. "Return-to-Sensibility Problems After Penetrating Captive Bolt Stunning of Cattle in Commercial Beef Slaughter Plants." *Journal of the American Veterinary Medical Association* 221 (2002): 1258–61.

Grazian, David. *American Zoo: A Sociological Safari*. Princeton, NJ: Princeton University Press, 2015.

Gregory, N. G., and L. J. Wilkins. "Broken Bones in Domestic Fowl: Handling and Processing Damage in End-of-Lay Battery Hens." *British Poultry Science* 30 (1989): 555–62.

Grimaldi, Wray W., et al. "Infectious Diseases of Antarctic Penguins: Current Status and Future Threats." *Polar Biology* 38 (2015): 591–606.

Gruen, Lori, ed. *The Ethics of Captivity*. New York: Oxford University Press, 2014.

Gunderson, Alex, and Jonathon Stillman. "Plasticity in Thermal Tolerance Has Limited Potential to Buffer Ectotherms from Global Warming." *Proceedings of the Royal Society B* (2015). doi:10.1098/rspb.2015.0401.

Hancocks, David. *A Different Nature: The Paradoxical World of Zoos and Their Uncertain Future*. Berkeley: University of California Press, 2001.

Hare, Brian, and Vanessa Woods. *The Genius of Dogs: How Dogs Are Smarter Than You Think*. New York: Plume, 2013.

Hargrove, John, and Howard Chua-Eoan. *Beneath the Surface: Killer Whales, SeaWorld, and the Truth Beyond Blackfish*. New York: St. Martin's, 2015.

Harris, Christine R., and Caroline Prouvost. "Jealousy in Dogs." *PLOS One* 9 (2014): e94597. doi:10.1371/journal.pone.0094597.

Harrison, Ruth. *Animal Machines: The New Factory Farming Industry*. Wallingford, UK: CABI, 2013. First published in 1964 by Vincent Stuart Publishers.

Haskell, M. J., et al. "Housing System, Milk Production, and Zero-Grazing Effects on Lameness and Leg Injury in Dairy Cows." *Journal of Dairy Science* 89 (2006): 4259–66.

Hastie, Gordon, et al. "Sound Exposure in Harbour Seals During the Installation of an Offshore Wind Farm: Predictions of Auditory Damage." *Journal of Applied Ecology* 52 (2015): 631–40.

Hawkins, P., et al. "Newcastle Consensus Meeting on Carbon Dioxide Euthanasia of Laboratory Animals." 2006. https://www.nc3rs.org.uk/sites/default/files/documents/Events/First%20Newcastle%20consensus%20meeting%20report.pdf.

Haynes, Richard P. *Animal Welfare: Competing Conceptions and Their Ethical Implications*. New York: Springer, 2008.

Hedgespeth, M. L., P. A. Nilsson, and O. Berglund. "Ecological Implications of Altered Fish Foraging After Exposure to an Antidepressant Pharmaceutical." *Aquatic Toxicology* 151 (2014): 84–87.

Hedrich, Hans J., ed. *The Laboratory Mouse*. 2nd ed. Waltham, MA: Academic Press, 2012.

Hernádi, Anna, et al. "Man's Underground Best Friend: Domestic Ferrets, Unlike the Wild Forms, Show Evidence of Dog-Like Social-Cognitive Skills." *PLOS One* 7 (2012): e43267. doi:10.1371/journal.pone.0043267.

Hervieux, Dave, et al. "Managing Wolves (*Canis lupus*) to Recover Threatened Woodland Caribou (*Rangifer tarandus caribou*) in Alberta." *Canadian Journal of Zoology* 92, no. 12 (2014). http://www.nrcresearchpress.com/doi/full/10.1139/cjz-2014-0142?src=recsys&#.V17QmFc4n-Y.

Hildebrandt, Isabel J., Helen Su, and Wolfgang A. Weber. "Anesthesia and Other

Considerations for in Vivo Imaging of Small Animals." *ILAR Journal* 49 (2008): 17–26.

Horowitz, Alexandra. "*Canis familiaris*: Companion and Captive." In *The Ethics of Captivity*, edited by Lori Gruen, 7–21. New York: Oxford University Press, 2014.

———. *Inside of a Dog: What Dogs See, Smell, and Know.* New York: Scribner, 2010.

Hosey, Geoff, Vicky Melfi, and Sheila Pankhurst. *Zoo Animals: Behaviour, Management, and Welfare.* New York: Oxford University Press, 2009.

Huang-Brown, K. M., and F. A. Guhad. "Chocolate, an Effective Means of Oral Drug Delivery in Rats." *Laboratory Animal* 31 (2002): 34–36.

Hubrecht, Robert C. *The Welfare of Animals Used in Research.* Ames, IA: Wiley Blackwell, 2014.

Hughes, B. O., and A. J. Black. "The Preference of Domestic Hens for Different Types of Battery Cage Floor." *British Poultry Science* 14 (1973): 615–19.

Humane Society of the United States. *An HSUS Report: The Welfare of Intensively Confined Animals in Battery Cages, Gestation Crates, and Veal Crates.* July 2012.

Jahng J. W., et al. "Hyperphagia and Depression-Like Behavior by Adolescence Social Isolation in Female Rats." *International Journal of Developmental Neuroscience* 30 (2012): 47–53.

Jamieson, Dale. "Against Zoos." In *Morality's Progress: Essays on Humans, Other Animals, and the Rest of Nature*, 166–75. Oxford, UK: Clarendon Press, 2002.

Jensen, Per. *The Ethology of Domestic Animals: An Introductory Text.* 2nd ed. Cambridge, MA: CABI, 2009.

Johnson, Steven. *Where Good Ideas Come From: The Natural History of Innovation.* New York: Riverhead Books, 2010.

Jones, Robert C. "Science, Sentience, and Animal Welfare." *Biology and Philosophy* 28 (2013): 1–30.

Julius, Henri, et al. *Attachment to Pets: An Integrative View of Human-Animal Relationships with Implications for Therapeutic Practice.* Boston: Hogrefe, 2012.

Kagan, Ron, Stephanie Allard, and Scott Carter. "Exotic Animal Welfare: A Path Forward." *WAZA Magazine* 16 (2015): 26.

Kagan, Ron, and Jake Veasey. "Challenges of Zoo Animal Welfare." In *Wild Mammals in Captivity*, edited by Devra G. Kleiman, Katerina V. Thompson, and Charlotte Kirk Baer, 11–21. Chicago: University of Chicago Press, 1996.

Kaufman, Leslie. "When Babies Don't Fit Plan, Question for Zoos Is, Now What?" *New York Times*, August 2, 2012. http://www.nytimes.com/2012/08/03/science/zoos-divide-over-contraception-and-euthanasia-for-animals.html.

Kellert, Stephen R. *Kinship to Mastery: Biophilia in Human Evolution and Development.* Washington, DC: Island Press, 1997.

Ketcham, Christopher. "The Rogue Agency." *Harper's Magazine.* March 2016, 38–44.

Kidd, K. A., et al. "Direct and Indirect Responses of a Freshwater Food Web to a Potent Synthetic Oestrogen." *Philosophical Transactions of the Royal Society B.* 369 (2014). doi:10.1098/rstb.2013.0578.

King, Barbara J. *How Animals Grieve.* Chicago: University of Chicago Press, 2013.

Kirby, David. *Death at Seaworld: Shamu and the Dark Side of Killer Whales in Captivity.* New York: St. Martin's Griffin, 2012.

Kirkden, Richard D., and Edmond A. Pajor. "Using Preference, Motivation and Aversion Tests to Ask Scientific Questions About Animals' Feelings." *Applied Animal Behaviour Science* 100 (2006): 29–47.

Knowles, Toby G., et al. "Leg Disorders in Broiler Chickens: Prevalence, Risk Factors and Prevention." *PLOS One* 3 (2006):e1545. doi:10.1371/journal.pone.0001545.

Kurtycz, L. M., K. E. Wagner, and S. R. Ross. "The Choice to Access Outdoor Areas Affects the Behavior of Great Apes." *Journal of Applied Animal Welfare Science* 17 (2014): 185–97.

Langbein, Jan, Katrin Siebert, and Gerd Nürnberg. "On the Use of an Automated Learning Device by Group-Housed Dwarf Goats: Do Goats Seek Cognitive Challenges?" *Applied Animal Behaviour Science* 120, nos. 3–4 (2009): 150–58.

Lay, D. C., et al. "Hen Welfare in Different Housing Systems." *Poultry Science* 90 (2011): 1–17.

Lewis, M. H., et al. "The Neurobiology of Stereotypy 1: Environmental Complexity." In Mason and Rushen, *Stereotypic Animal Behaviour*, 190–226.

Lorimer, Jamie. *Wildlife in the Anthropocene: Conservation After Nature.* Minneapolis: University of Minnesota Press, 2015.

Lürzel, Stephanie, et al. "The Influence of Gentle Interactions on Avoidance Distance Towards Humans, Weight Gain and Physiological Parameters in Group-Housed Dairy Calves." *Applied Animal Behaviour Science* 172 (2015): 9–16.

Lusk, Jayson L. "The Market for Animal Welfare." *Agriculture and Human Values* 28 (2011): 561–75.

Makowska, Joanna I., and Daniel M. Weary. "Assessing the Emotions of Laboratory Rats." *Applied Animal Behaviour Science* 148, nos. 1–2 (2013): 1–12.

Mallonée, Jay S., and Paul Joslin. "Traumatic Stress Disorder Observed in an Adult Wild Captive Wolf (*Canis lupus*)." *Journal of Applied Animal Welfare Science* 7 (2004): 107–26.

Margodt, Koen. *The Welfare Ark: Suggestions for a Renewed Policy for Zoos.* Amsterdam: VU University Press, 2001.

Markowitz, Hal, and Gregory Timmel. "Animal Well-Being and Research Outcomes." In *Mental Health and Well-Being in Animals*, edited by Frank McMillan, 277–83. Ames, IA: Blackwell, 2005.

Markowitz, Hal, and Katherine Eckert. "Giving Power to Animals." In *Mental Health and Well-Being in Animals*, edited by Frank McMillan, 201–09. Ames, IA: Blackwell, 2005.

Markowitz, Hal. *Enriching Animal Lives.* Pacifica, CA: Mauka Press, 2011.

Marks, Joel. "A Is for Animal: The Animal User's Lexicon." *Between the Species* 18, no. 1 (August 2015).

Martin, Bronwen, et al. "'Control' Laboratory Rodents Are Metabolically Morbid: Why It Matters." *Proceedings of the National Academy of Sciences* 107 (2010): 6127–33. doi:10.1073/pnas.0912955107.

Martin, Paul, and Patrick Bateson. *Measuring Behavior: An Introductory Guide.* New York: Cambridge University Press, 2007.

Mason, Georgia, and Jeffrey Rushen, eds. *Stereotypic Animal Behaviour: Fundamentals and Application to Welfare.* 2nd ed. Wallingford, UK: CABI, 2008.

Mason, Georgia. "Animal Welfare: The Physiology of the Hunted Deer." *Nature* 391 (1997): 22.

———. "Stereotypies: A Critical Review." *Animal Behaviour* 41 (1991): 1015–37.

———. "Using Species Differences in Health and Well-Being to Identify Intrinsic Risk and Protective Factors." *World Association of Zoos and Aquarium Magazine* (2015): 2–5.

McCowan, B., et al. "Effects of Induced Molting on the Well-Being of Egg-Laying Hens." *Journal of Applied Animal Welfare Science* 9 (2006): 9–23.

McDonald, M. A., J. A. Hildebrand, and S. M. Wiggins. "Increases in Deep Ocean Ambient Noise in the Northwest Pacific West of San Nicolas Island, California." *Journal of the Acoustical Society of America* 120, no. 2 (2006): 711–18.

McGowan, Ragen T. S., et al. "Positive Affect and Learning: Exploring the 'Eureka Effect' in Dogs." *Animal Cognition* 17 (2014): 577–87.

McLeod, Steven, and Trudy Sharp. "Improving the Humaneness of Commercial Kangaroo Harvesting." Kingston, Australia: Rural Industries Research and Development Corporation, 2014.

McMillan, Frank. "Emotional Maltreatment in Animals." In *Mental Health and Well-Being in Animals*, edited by Frank McMillan, 167–77. Ames, IA: Blackwell, 2005.

———. "Stress-Induced and Emotional Eating in Animals: A Review of the Experimental Evidence and Implications for Companion Animal Obesity." *Journal of Veterinary Behavior* 8 (2013): 376–85.

McWilliams, James. *The Modern Savage*. New York: Thomas Dunne, 2015.

Mehrkam, Lindsay, and Nicole Dorey. "Is Preference a Predictor of Enrichment Efficacy in Galapagos Tortoises (*Chelonoidis Nigra*)?" *Zoo Biology* 33 (2014): 275–84.

Melfi, V. A. "There Are Big Gaps in Our Knowledge, and Thus Approach, to Zoo Animal Welfare: A Case for Evidence-Based Zoo Animal Management." *Zoo Biology* 28 (2009): 574–88.

Mellor, David J. "Updating Animal Welfare Thinking: Moving Beyond the 'Five Freedoms' Toward 'A Life Worth Living,'" *Animals* 6, no. 3 (2016): 21.

Mellor, David J., and N. J Beausoleil. "Extending the 'Five Domains' Model for Animal Welfare Assessment to Incorporate Positive Welfare States." *Animal Welfare* 24 (2015): 214–53.

Mellor, David, Emily Patterson-Kane, and Kevin Stafford. *The Sciences of Animal Welfare*. Ames, IA: Wiley-Blackwell, 2009.

Mench, J., and J. Swanson. "Developing Science-Based Animal Welfare Guidelines." Paper presented at University of California Poultry Symposium and Egg Processing Workshop. Modesto, CA, November 7, 2000. Available at http://animalscience.ucdavis.edu/avian/mench.pdf.

Mepham, Ben. "Ethical Analysis of Food Biotechnologies: An Evaluative Framework." In *Food Ethics*, edited by Ben Mepham, 101–19. New York: Routledge, 1996.

Miklosi, Adam. *Dog Behavior, Evolution, and Cognition*. 2nd ed. New York: Oxford University Press, 2015.

Millar, Kate M. "Respect for Animal Autonomy in Bioethical Analysis: The Case of Automated Milking Systems (AMS)." *Journal of Agriculture and Environmental Ethics* 12, no. 1 (2000): 41–50.

Miller, Lance J., et al. "Animal Welfare Management of Bottlenose Dolphins at the Chicago Zoological Society's Brookfield Zoo." *World Association of Zoos and Aquarium Magazine* (2015): 14–17.

Minteer, Ben A., et al. "Avoiding (Re)extinction." *Science* 344 (2014): 260–61.

Monamy, Vaughan. *Animal Experimentation: A Guide to the Issues*. New York: Cambridge University Press, 2000.

Monbiot, George. *Feral: Rewilding the Land, the Sea, and Human Life*. Chicago: University of Chicago Press, 2015.

Mounier, L. "Pre-slaughter Conditions, Animal Stress and Welfare: Current Status and Possible Future Research." *Animal* 2 (2008): 1501–17.

Murphy, E., R. E. Nordquist, and F. J. van der Staay. "A Review of Behavioural Methods to Study Emotion and Mood in Pigs, *Sus scrofa*." *Applied Animal Behaviour Science* 159 (2014): 9–28.

Nagy, Kelsi, and Phillip David Johnson II, eds. *Trash Animals: How We Live with Nature's Filthy, Feral, Invasive, and Unwanted Species*. Minneapolis: University of Minnesota Press, 2013.

Neave, Heather W., et al. "Pain and Pessimism: Dairy Calves Exhibit Negative Judgement Bias Following Hot-Iron Disbudding." *PLOS One* 8 (2013): e80556. doi:10.1371/journal.pone.0080556.

Nibert, David A. *Animal Oppression and Human Violence: Domesecration, Capitalism, and Global Conflict*. New York: Columbia University Press, 2013.

Normando, Simona, and Donatella Gelli. "Behavioral Complaints and Owners' Satisfaction in Rabbits, Mustelids, and Rodents Kept as Pets." *Journal of Veterinary Behavior* 6 (2011): 337–42.

Norwood, Bailey, and Jayson Lusk. *Compassion, by the Pound: The Economics of Farm Animal Welfare*. New York: Oxford University Press, 2011.

Noske, Barbara. *Humans and Other Animals: Beyond the Boundaries of Anthropology*. London: Pluto Press, 1989.

Oaks, J. Lindsay, et al. "Diclofenac Residues as the Cause of Vulture Population Decline in Pakistan." *Nature* 427 (2004): 630–33.

Olmert, Meg. *Made for Each Other: The Biology of the Human-Animal Bond*. Boston: Da Capo, 2010.

Ordiz, Andrés, et al. "Do Bears Know They Are Being Hunted?" *Biological Conservation* 152 (2012): 21–28.

Orleans, F. Barbara. *In the Name of Science: Issues in Responsible Animal Experimentation*. New York: Oxford University Press, 1993.

Owen, Megan A., et al. "Hearing Sensitivity in Context: Conservation Implications for a Highly Vocal Endangered Species." *Global Ecology and Conservation* 6 (2016): 121–31.

Pacelle, Wayne. *The Humane Economy: How Innovators and Enlightened Consumers Are Transforming the Lives of Animals*. New York: William Morrow, 2016.

Palmer, Clare, and Peter Sandøe. "For Their Own Good: Captive Cats and Routine Confinement." In *The Ethics of Captivity*, edited by Lori Gruen, 135–55. New York: Oxford University Press, 2014.

Panksepp, Jaak. "Science of the Brain as a Gateway to Understanding Play." Interview with Jaak Panksepp. *American Journal of Play* 3 (2010): 245–77.

Patterson-Kane, E. G., D. N. Harper, and M. Hunt. "The Cage Preferences of Laboratory Rats." *Laboratory Animals* 35 (2001): 74–79.

Pearce, Fred. *The New Wild: Why Invasive Species Will Be Nature's Salvation.* Boston: Beacon Press, 2015.

Peterson, Dale. *Where Have All the Animals Gone? My Travels with Karl Ammann.* Peterborough, NH: Bauhan, 2015.

Pierce, Jessica. *The Last Walk: Reflections on Our Pets at the Ends of Their Lives.* Chicago: University of Chicago Press, 2012.

———. *Run, Spot, Run: The Ethics of Keeping Pets.* Chicago: University of Chicago Press, 2016.

Preilowski, Bruno, Michael Reger, and Hans Engele. "Combining Scientific Experimentation with Conventional Housing: A Pilot Study with Rhesus Monkeys." *American Journal of Primatology* 14 (1988): 223–34.

Proctor, Helen S., and Gemma Carder "Can Ear Posture Reliably Measure the Positive Emotional State of Cows?" *Applied Animal Behaviour Science* 161 (2014): 20–27.

———. "Nasal Temperatures in Dairy Cows Are Influenced by Positive Emotional State." *Physiology & Behavior* 138 (2015): 340–44.

Proctor, Helen S., Gemma Carder, and Amelia R. Cornish. "Searching for Animal Sentience: A Systematic Review of the Scientific Literature." *Animals* 3 (2013): 882–906.

Proulx, Gilbert, and Dwight Rodtka. "Predator Bounties in Western Canada Cause Animal Suffering and Compromise Wildlife Conservation Efforts." *Animals* 5 (2015): 1034–46.

Proulx, Gilbert, et al. "Humaneness and Selectivity of Killing Neck Snares Used to Capture Canids in Canada: A Review." *Canadian Wildlife Biology & Management* 4 (2015): 55–65.

———. "Poisoning Wolves with Strychnine Is Unacceptable in Experimental Studies and Conservation Programmes." *Environmental Conservation* (2015). doi:10.1017 /S0376892915000211.

Raman, T. R. Shankar. "Leopard Landscapes: Coexisting with Carnivores in Countryside and City." *Economy and Political Weekly* 50, no. 1, January 3, 2015. http://www .epw.in/journal/2015/1/reports-states-web-exclusives/leopard-landscapes.html-0.

Ramp, Daniel, and Marc Bekoff. "Compassion as a Practical and Evolved Ethic for Conservation." *Bioscience* (2015): 1–5.

Reefman, Nadine, et al. "Ear and Tail Postures as Indicators of Emotional Valence in Sheep." *Applied Animal Behaviour Science* 118 (2009): 199–207.

Rees, Paul A. *An Introduction to Zoo Biology and Management.* Hoboken, NJ: Wiley-Blackwell, 2011.

Regan, Tom. *The Case for Animal Rights.* Berkeley: University of California Press, 1983.

Reimert, I., et al. "Emotions on the Loose: Emotional Contagion and the Role of Oxytocin in Pigs." *Animal Cognition* 18 (2015): 517–32.

Reinhardt, Viktor, and Annie Reinhardt. *Variables, Refinement and Environmental Enrichment for Rodents and Rabbits Kept in Research Institutions: Making Life Easier for Animals in Laboratories.* Washington, DC: Animal Welfare Institute, 2006.

Robertson, I. *Animals, Welfare and the Law: Fundamental Principles for Critical Assessment.* New York: Routledge, 2015.

Rooney, Nicola, and John Bradshaw. "Canine Welfare Science: An Antidote to Sentiment and Myth." In *Domestic Dog Cognition and Behavior*, edited by Alexandra Horowitz. Berlin: Springer-Verlag, 2014.

Rudacille, Deborah. *The Scalpel and the Butterfly: The War Between Animal Research and Animal Protection*. New York: Farrar, Straus and Giroux, 2000.

Rushen, J. "Assessing the Welfare of Dairy Cattle." *Journal of Applied Animal Welfare Science* 4 (2001): 223–34.

Russell, William, and Rex Burch. *The Principles of Humane Experimental Technique*. Herts, England: Universities Federation for Animal Welfare, 1959.

Ryder, Richard. *Animal Revolution: Changing Attitudes Toward Speciesism*. New York: Bloomsbury Academic, 1989.

Safina, Carl. *Beyond Words: What Animals Think and Feel*. New York: Henry Holt, 2015.

Salvanes, Anne Gro Vea, et al., "Environmental Enrichment Promotes Neural Plasticity and Cognitive Ability in Fish." *Proceedings of the Royal Society B*, 280 (2013): 1767.

Sandem, A. I., and B. O. Braastad. "Effects of Cow–Calf Separation on Visible Eye White and Behaviour in Dairy Cows—A Brief Report." *Applied Animal Behaviour Science* 95 (2005): 233–39.

Sandem, A. I., B. O. Braastad, and K. E. Bøe. "Eye White May Indicate Emotional State on a Frustration-Contentedness Axis in Dairy Cows." *Applied Animal Behaviour Science* 79 (2002): 1–10.

Sato, Nobuya, et al. "Rats Demonstrate Helping Behavior Toward a Soaked Conspecific," *Animal Cognition* 18 (2015): 1039–47.

Savage-Rumbaugh, Sue, et al. "Welfare of Apes in Captive Environments: Comments On, and By, a Specific Group of Apes." *Journal of Applied Animal Welfare Science* 10, no. 1 (2007): 7–19.

Seegers, H., C. Fourichon, and F. Beaudeau. "Production Effects Related to Mastitis and Mastitis Economics in Dairy Cattle Herds." *Veterinary Research* 34 (2003): 475–91.

Seligman, M. E. P., S. F. Maier, and J. H. Geer. "Alleviation of Learned Helplessness in the Dog." *Journal of Abnormal Psychology* 73 (1968): 256–62.

Shamoo, Adil E., and David B. Resnik. *Responsible Conduct of Research*. New York: Oxford University Press, 2003.

Shepherdson, David, et al. "Individual and Environmental Factors Associated with Stereotypic Behavior and Fecal Glucocorticoid Metabolite Levels in Zoo Housed Polar Bears." *Applied Animal Behaviour Science* 147 (2013): 268–77.

Sherwen, Sally L., et al. "Little Penguins, *Eudyptula minor*, Show Increased Avoidance, Aggression and Vigilance in Response to Zoo Visitors." *Applied Animal Behaviour Science* 168 (2015): 71–76.

Siegford, Janice M. "Multidisciplinary Approaches and Assessment Techniques to Better Understand and Enhance Zoo Nonhuman Animal Welfare." *Journal of Applied Animal Welfare Science* 16 (2013): 300–318.

Simpson, Stephen D., et al. "Anthropogenic Noise Increases Fish Mortality by Predation," *Nature Communications* (February 2016). doi:10.1038/ncmms10544.

Singer, Peter. *Animal Liberation*, 2nd ed. New York: HarperCollins, 2001.

Snowdon, Charles T., David Teie, and Megan Savage. "Cats Prefer Species-Appropriate Music." *Applied Animal Behaviour Science* 166 (2015): 106–11.

Sonntag, Q., and K. L. Overall. "Key Determinants of Dog and Cat Welfare: Behaviour, Breeding, and Household Lifestyle." *Scientific and Technical Review of the Office International des Epizooties* (Paris) 33 (2014): 213–20.

Špinka, M., R. C. Newberry, and M. Bekoff. "Mammalian Play: Training for the Unexpected." *Quarterly Review of Biology* 76 (2001): 141–68.

Špinka, Marek, and Françoise Wemelsfelder. "Environmental Challenge and Animal Agency." In Appleby and Hughes, *Animal Welfare*, 27–43.

Stafford, K., and D. Mellor. "The Welfare Significance of the Castration of Cattle: A Review." *New Zealand Veterinary Journal* 53 (2005): 271–78.

Starling, Melissa, et al. "Canine Sense and Sensibility: Tipping Points and Response Latency Variability as an Optimism Index in a Canine Judgement Bias Assessment." *PLOS One* 9 (2014). doi:10.1371/journal.pone.0107794.

Sterelny, Kim. *The Evolution of Agency and Other Essays.* New York: Cambridge University Press, 2001.

Sutherland, M. A., and C. B. Tucker. "The Long and Short of It: A Review of Tail Docking in Farm Animals." *Applied Animal Behaviour Science* 135 (2011): 179–91.

Suzuki, T., et al. "Preferences for Opioids by the Weight Pulling Method in Rats." *Pharmacology and Biochemical Behavior* 35 (1990): 413–18.

Svartberg, Kenth, et al. "Consistency of Personality Traits in Dogs." *Animal Behaviour* 69 (2005): 283–91.

Swaisgood, Ron, and David Shepherdson. "Environmental Enrichment as a Strategy for Mitigating Stereotypies in Zoo Animals." In Mason and Rushen, *Stereotypic Animal Behaviour*, 256–85.

Szentágothai, J. "The 'Brain-Mind' Relation: A Pseudo-Problem?" In *Mindwaves: Thoughts on Intelligence, Identity and Consciousness*, edited by C. Blakemore and S. Greenfield, 323–36. Oxford, UK: Basil Blackwell, 1987.

Taylor, A. A., and D. M. Weary. "Vocal Responses of Piglets to Castration: Identifying Procedural Sources of Pain." *Applied Animal Behaviour Science* 70 (2000): 17–26.

Taylor, Paul. *Respect for Nature: A Theory of Environmental Ethics.* Princeton, NJ: Princeton University Press, 1986.

Teixeira, Camila, et al. "Revisiting Translocation and Reintroduction Programmes." *Animal Behavior* 73 (2006): 1–13.

Townsend, P. "Use of In-Cage Shelters by Laboratory Rats." *Animal Welfare* 6 (1997): 95–103.

Turner, Dennis C., and Patrick Bateson, eds. *The Domestic Cat: The Biology of Its Behaviour.* 3rd ed. New York: Cambridge University Press, 2014.

Tuttle, Merlin. *The Secret Life of Bats: My Adventures with the World's Most Misunderstood Mammals.* New York: Houghton Mifflin Harcourt, 2015.

Vallortigara, G., et al. "Are Animals Autistic Savants?" *PLoS Biology* 6 (2008): e42. doi:10.1371/journal.pbio.0060042.

Van Dooren, Thom. "Invasive Species in Penguin Worlds: An Ethical Taxonomy of Killing for Conservation." *Conservation and Society* 9 (2011): 286–98.

Vickers, T. Winston, et al. "Survival and Mortality of Pumas (*Puma concolor*) in a Fragmented, Urbanizing Landscape." *PLOS One* (July 15, 2015). http://dx.doi .org/10.1371/journal.pone.0131490.

Visak, Tatjana, and Robert Garner, eds. *The Ethics of Killing Animals*. New York: Oxford University Press, 2016.

Waiblinger, Susanne, et al. "Assessing the Human-Animal Relationship in Farmed Species: A Critical Review." *Applied Animal Behaviour Science* 101 (2006): 185–242.

Waldau, Paul. *Animal Studies: An Introduction*. New York: Oxford University Press, 2013.

Warriss, P. D. "The Handling of Cattle Pre-slaughter and Its Effects on Carcass and Meat Quality." *Applied Animal Behaviour Science* 28 (1990): 171–86.

———. "The Welfare of Slaughter Pigs During Transport." *Animal Welfare* 7 (1998): 365–81.

Warwick, Clifford. "The Morality of the Reptile 'Pet' Trade." *Journal of Animal Ethics* 4 (2014): 74–94.

Watson, Lyall. *The Whole Hog: Exploring the Extraordinary Potential of Pigs*. London: Profile Books, 2004.

Webster, John. *Animal Welfare: A Cool Eye Toward Eden*. Hoboken, NJ: Wiley-Blackwell, 1995.

———, ed. *Management and Welfare of Farm Animals: The UFAW Farm Handbook*. Ames, IA: Wiley Blackwell, 2011.

Weeks, C. A., et al. "The Behaviour of Broiler Chickens and Its Modification by Lameness." *Applied Animal Behaviour Science* 67 (2000): 111–25.

Wemelsfelder, Françoise, et al. "Assessing the 'Whole Animal': A Free Choice Profiling Approach." *Animal Behaviour* 62 (2001): 209–20.

Wemelsfelder, Françoise. "Animal Boredom: Understanding the Tedium of Confined Lives." In *Mental Health and Well-Being in Animals*, edited by Frank McMillan, 79–92. Oxford, UK: Blackwell, 2005.

Wendler, David. "Should Protections for Research with Humans Who Cannot Consent Apply to Research with Nonhuman Primates?" *Theoretical Medicine and Bioethics* 35 (2014): 157–73.

Wilson, Lisa, and Ann Baldwin. "Effects of Environmental Stress on the Architecture and Permeability of the Rat Mesenteric Microvasculature." *Microcirculation* 5, no. 4 (1998): 299–308.

———, "Environmental Stress Causes Mast Cell Degranulation, Endothelial and Epithelial Changes, and Edema in the Rat Intestinal Mucosa." *Microcirculation* 6, no. 3 (1999): 189–98.

Wolfson, David J. "Beyond the Law: Agribusiness and the Systemic Abuse of Animals." *Lewis and Clark Law Review* (1996).

World Animal Protection. "Discover Sentience Mosaic." http://www.worldanimal protection.org/our-work/education-animal-welfare/discover-sentience-mosaic.

Würbel, H. "The Motivational Basis of Caged Rodents' Stereotypies." In Mason and Rushen, *Stereotypic Animal Behaviour*, 86–120.

Yin, Sophia. *How to Behave So Your Dog Behaves*. 2nd ed. Neptune, NJ: TFH, 2010.

Young, Robert. *Environmental Enrichment for Captive Animals.* Ames, IA: Blackwell, 2003.

Zonderland, J. J., et al. "Measuring a Pig's Preference for Suspended Toys by Using an Automated Recording Technique." *Agricultural Engineering International: The CIGR Journal of Scientific Research and Development* (2003): 1–12. https://ecommons.cornell.edu/handle/1813/10310.

Zulkifli, I., and A. Siti Khatijah. "The Relationships Between Cage Floor Preferences and Performance in Broiler Chickens." *Asian-Australasian Journal of Animal Sciences* 11, no. 3 (March 1998): 234–38.

Index

prevalence of euthanasia at, 41, 120,
127, 129. *See also* companion animals;
Humane Society of the United States
(HSUS); pet industry
"Animal User's Lexicon" (Marks), 50
animal welfare: defining, 18–22, 27–28;
Five Freedoms on, 4–7, 10, 17, 27, 33,
37; vs. productivity, 33–34, 46–50,
60; vs. well-being, 7, 23–24, 173–77,
179–80. *See also* animal welfare science; animal well-being
Animal Welfare Act (U.S., 1966), 16, 64,
67–68, 187n14
"animal welfare approved," 51
animal welfare organizations. See *specific
organization names*
animal welfare science: on companion
animals, 22–23, 120–21; defining
welfare, 16–22, 23, 27–28; development and expansion of, 22–23; on
enrichment and stereotypies, 104;
ethics of, 137–38, 173–77; on food animals, 4–5, 33–34; future of, 171–73;
language of emotions in, 135–36;
limitations of, 23–28, 62–64, 171,
173–74; literature on suffering, 39–40;
vs. meat science, 48–50; origins of, 10,
32, 62; preference testing, 17, 19–20,
22, 40, 56; rise in popularity of, 16–18;
vs. science of well-being, 7, 23–24;
on trapping, 145–46; truisms in, 10,
52–53, 183n1; use of term humane in,
50–52; on wild animals, 140; on zoo
animals, 94–98, 109–110, 115–16. *See
also* animal welfare; scientific animal
research
Animal Welfare Science and Bioethics
Centre, 171–72
animal well-being: definition of, 7; vs.
human well-being, 82–83; individual
focus of, 55–58; science of, 28–30, 116,
164–67; vs. welfare, 7, 23–24, 173–77,
179–80. *See also* animal welfare;
animal welfare science; freedom
Antarctica, 161
Anthropocene, as term, 7, 167

Antwerp Zoo, 101
apes, 78, 82–85, 88–89, 188n53. *See also*
chimpanzees; monkeys
aquariums, 2, 14, 16, 91–92. *See also*
captivity; zoos
Arctic foxes, 151
Argentina, 171
ARK Sanctuary, 114–15
arthritis, 100
"The Assessment of Pain and Distress in
Animals" (Thorpe), 17
Association of Zoos and Aquariums
(AZA), 94, 96, 111
Atlantic, 179
augmented reality in zoos, 112
Australia, 148–49, 165
Austria, 84
automated milking systems (AMS), 54
autonomy, 81–82
aversion research, 22. *See also* preference
testing

baboons, 165
bamboo pit vipers, 150–51
Barber, Jesse, 159
barred owls, 148, 152–53
Barred Owl Stakeholder Group, 152, 153
Bateson, Patrick, 129–30, 141
bats, 148–49, 159
battery cages, 4, 17, 19, 35–36, 39–40, 60.
See also cage-free chickens; chickens
Baumans, Vera, 70
Bear 317, 144, 166
bearded dragons, 136
bears. See *specific types*
Beauchamp, Tom, 81
bee fly, 156
behavioral freedom: of chickens, 34–37,
53; of cows, 37–38; of food animals,
34–38, 53–55; science experiments
and, 86–87; voluntariness, 54, 78–81;
of wild animals, 102, 103. *See also*
freedom; social behaviors; stereotypic
behaviors; stress
Beijing Zoo, 105
Bekoff, Marc: on advocacy and educa-